SUPERCAT

Raising the Perfect Feline Companion

History's first Supercat probably emerged in ancient Egypt where cats were venerated and the cat-headed goddess "Bast, the Lady of Life" was one of the society's most revered deities. Statues like this one, paying homage to the cat, have been found all over the country.

Photo courtesy M. W. Fox

SUPERCAT

Raising the Perfect Feline Companion

Dr. Michael W. Fox

HOWELL
BOOK HOUSE

New York

Maxwell Macmillan Canada
Toronto

Maxwell Macmillan International
New York   Oxford   Singapore   Sydney

Howell Book House
Macmillan Publishing Company
866 Third Avenue, New York, NY 10022

Maxwell Macmillan Canada, Inc.
1200 Eglinton Avenue East, Suite 200
Don Mills, Ontario M3C 3N1

Macmillan Publishing Company is part of the Maxwell Communication Group of
Companies.

Library of Congress Cataloging-in-Publication Data
Fox, Michael W., 1937–
    Supercat : raising the perfect feline companion / by Michael W. Fox.
       p.   cm.
    Includes index.
    ISBN 0-87605-843-8
    1. Cats. 2. Cats—Behavior. I. Title.
SF442.F68 1990
636.8'088'7—dc20   90-15526   CIP

Macmillan books are available at special discounts for bulk purchases for sales
promotions, premiums, fund-raising, or educational use. For details, contact:

                        Special Sales Director
                        Macmillan Publishing Company
                        866 Third Avenue
                        New York, NY 10022

10   9   8   7   6   5   4   3   2   1

Printed in the United States of America

For all cats wild and tame.

In Egypt, when a cat died, it was common practice to mummify the remains. These remarkable examples in the British Museum show the esteem in which these ancients held their cats.

Photo courtesy M. W. Fox

Contents

The inborn potential of intelligence in all cats is affected by social and environmental experiences in their early lives. It is the responsibility of their human custodians to provide the correct upbringing of young cats to make these experiences as rich and fulfilling as possible. *Photo courtesy HSUS/Botnovcan*

Preface

AS I WILL DETAIL in this book, if our animal companions are not raised to be trusting and somewhat dependent upon their human companions and are not given the right social and environmental experiences early in life, then their intelligence and trainability will suffer. I regard intelligence as an inborn potential in all animals, a potential that can be inhibited through improper rearing and severely limited by improper breeding. Trainability—or educability—depends upon animals' having good IQ or intelligence potential plus a strong emotional and motivational connection with their human companions. Thus we humans play an important role in determining how responsive and intelligent our animal companions are. And the more responsive and intelligent they are, the more satisfying they are as companions, and the more their potential can be fulfilled—which could be considered a basic "right" of all domesticated animals.

Today we have a long and growing list of cat shows in many areas, but as the name implies, the focus is more upon physical appearance than on developing our cats' potentials and exploring their own natural conceptual abilities. My hope is that once IQ tests and pet potentials are more widely used and recognized, we will have a new era for petdom: a new age in which we do not simply train cats and other animals to do tricks but educate them in the full sense of the word.

When I conducted some of the tests described in this book on one of my cats who seemed to be a rather lazy and disinterested subject, he came out with top marks. He not only enjoyed "going to school," he also gained the esteem of other members of the family who thought he was a lazy layabout, myself included.

Older children and adults will enjoy taking their pets through the many tests and lessons in this book as well as learning how to raise their feline companions in order to maximize their intelligence and other inborn potentials. The outcome will inevitably lead to a greater respect and understanding for the animals in our lives—and a more entertaining and enriching life for the animals themselves.

It is evident that I have high aspirations for the positive impact SUPERCAT can make for companion cats everywhere. I also think these aspirations are entirely realistic, but no book of this kind is the total effort of one mind and one pair of hands. Many have helped make SUPERCAT what it is today. There are special acknowledgments to make at this point to my wife, Deanna Krantz, for her critical and sensitive editorial work and research in preparing this work, to Ellen Truong for her invaluable secretarial help in the preparation of the manuscript, to my editor, Seymour Weiss, for his continued interest and his particular abilities, and to the staff of Howell Book House for its ongoing, enthusiastic support.

—Dr. Michael W. Fox
Washington, D.C.

Introduction

THE QUESTIONS of human intelligence and IQ tests have long been controversial issues, spiced with sexist and racist overtones. And the question of intelligence is no less controversial when it comes to animals.

People often contend that cats must be less intelligent than dogs because they can't be taught tricks or be obedience trained. This attitude reflects the erroneous belief that the more trainable animals are, the more intelligent they must be. However, no matter how intelligent animals are, if they are not emotionally attached to their trainers, they will be extremely difficult to train. Hence the importance of proper rearing to develop a close emotional bond with the animal.

Cats are more independent and less socially dependent than dogs; hence they are more difficult to train, but that does not mean that they are less intelligent than dogs. In actuality, there's nothing more intelligent at being a cat than a cat.

It is surprising to me that people have long believed that animals are not really intelligent, that they are incapable of reason and that they are irrational and governed by "instinct," which must be controlled through discipline and trained through simple conditioning (usually aversive, that is, painful) if they are to be domesticated and kept as pets.

So I will first demonstrate that felines are not only sentient (have feelings) but are also intelligent (sapient). This is essential, I believe, before the right attitude can be established, which is needed if we are to raise and train our cats properly and enjoy the company of such intelligent animal companions. I will also describe ways to evaluate their intelligence and look into how their potentials can be enhanced by special rearing and enrichment procedures, especially early in life.

The ability of animals to reason is a concept that has intrigued man and precipitated discussion for centuries. Cultural conditioning has argued against the possibility, but there is an overwhelming amount of documentation to support it.   *Photo courtesy HSUS/Trena*

1

Do Animals Think?
Can They Reason?

D O ANIMALS REALLY THINK and show foresight and insight? Can they be rational and can they reason? There is a widespread belief that animals, with perhaps the exception of chimpanzees, can do none of these things. Several scientific studies of animals' behavior and learning abilities, and the experiences of pet owners with their cats and dogs, dispel this belief. Yet many people choose to hold on to this erroneous view of animals as unthinking and even unfeeling machines. This attitude has its historical roots in theology and philosophy, setting humans above animals so that we can exploit them without a twinge of conscience.

Now it is true that much of an animal's behavior is instinctive-mechanical, like a dog's wagging its tail or a cat's purring. Human laughter, smiling and crying are no less instinctive. While these actions are instinctual or inborn, when and to whom a cat purrs entails discrimination (as between a friend or foe). The ability to discriminate involves the rudiments of intelligent reasoning, more so when the cat purrs and head-rubs in order to influence, even manipulate, its owner for attention, for food or to be played with. A cat will even bring a toy to ask its owner to play with it. Such behavior is more than simple mechanical conditioning, since reasoning and insight underlie these complex symbolic actions, where the animal has expectations and can anticipate the outcome of its actions. An animal that could not think, make associations and see into the future in terms of the causal sequence of events could not perform such actions in the first place.

If animals can't think or reason, try to explain this cat's behavior. Felix lives in a New York City high rise. When he first moved in with his mistress, she fixed a bell on a string on the inside of the apartment door and another on the outside, showed Felix how to swipe the inside bell, led him out, pressed the elevator button and took him down to the ground floor. The doorman opened the main door, which led onto an enclosed lawn and garden. Felix explored for a while, then was taken back indoors, up in the elevator to the seventh floor, and shown how to hit the bell on his apartment door. After a few day's further help, Felix, through observational learning (a true "copycat"), mastered the procedure and now rings the bell when he wants to go in or out; he waits for someone to use the elevator to go down; the doorman helps him with the front door, and when Felix returns, presses the button and lets the cat enter the elevator. Felix never gets off on the wrong floor even if other people happen to be on the elevator.

While I do not endorse the practice of ever letting cats go outdoors unsupervised, the phenomenal Felix demonstrates the point that such advanced learning entails far more than thoughtless mechanical conditioning. But this doesn't mean that cats are always rational. Like us, they are often quite irrational and act without reason. A cat or human in a state of hysterical terror or paranoia cannot be reasoned with. As humans get carried away with their emotions and instinctive desires, so animals sometimes get out of control. For example, a cat may attack its companion cat because it smells of the veterinarian's hospital where it has just been; or a cat may even attack its owner when they meet outdoors if the cat is alarmed by the scent marks of feline intruders in the garden.

Cats are emotionally similar to us and have comparable, if less developed, mental abilities. It would be wrong, indeed irrationally anthropomorphic, to believe that animals have anywhere near our abilities to think abstractly and make intuitive associations.

While it takes highly sophisticated intelligence and technology for an aviator to plot a course from Alaska or Missouri to Peru or Mexico, migrant birds and monarch butterflies can do this instinctively. Their ability is inborn; ours is learned. They respond perhaps without objective awareness: their awareness is intuitive, no less re-

Cats send a variety of messages through unmistakable body language. This head rub is an affectionate gesture that also allows the cat to scent mark the person providing the petting. *Photo courtesy M. W. Fox*

Rolling, often combined with vocalization, is a play/petting solicitation. Most cat lovers will agree it's a signal that is very hard to resist. *Photo courtesy M. W. Fox*

markable than our acquired abilities and, in essence, no less intelligent.

This leads us to another related issue: are animals really consciously aware? Once we learn to do something, such as drive a car, our repetition of the action becomes unconscious, which explains my difficulty in regaining conscious attention so as to not jam the gears on my wife's new car, whose stick shift is patterned differently from that of my own car and her deceased one. So we aren't always consciously aware when we execute learned, habitual actions. Nor are animals that have learned some routine or are executing some complex instinctual behavior: we and they are creatures of habit.

Determining how consciously aware animals are is difficult, but the writings of two eminent scientific authorities, Professor Donald Griffin in his book *Animal Awareness* and Dr. Stephen Walker in *Animal Thought,* lead to the inevitable conclusion that animals are, to varying degrees, consciously aware and are capable of reason, insight and intelligent behavior. Clever creatures do think, and the more social and empathetic and emotive they are, the more they can think.

Certainly computers are intelligent, and many animal species, especially insects, show a mechanical, computerlike intelligence. While there is certainly some awareness—electronic or neurological— scientists are still uncertain which animal species possess self-awareness. A fish or bird will attack or court its image in a mirror, such primary narcissism being suggestive of a lack of self-awareness. But as a bird will preen itself when dirty and a fish can discriminate when it is being touched and when it is touching something itself, the basic ingredients of self-awareness and thus reflective thought must be present in these animals. Most kittens will react to a mirror as though it were another animal, but as they mature, they will ignore their mirror image, which implies that they know it is not another animal but their own reflection. So they seem to be self-aware.

Many people have a very narrow view of animal intelligence, often based upon the restricted opportunities of their pets and other animals (in zoos, laboratories and "factory" farms) to show their natural talents. Little wonder that such people deny that these animals have the ability to reason, and any learning of which their owners are aware is usually dismissed as simple conditioning. The ultimate measure of any species' intelligence is its ability to survive, and until we began to threaten the animal kingdom with extinction under our unfeeling dominion, we should remember that animal species have suc-

He knows it isn't another cat. Cats will show an interest in a reflected image, but are not tricked into thinking that what they are seeing is another animal.

Photo courtesy M. W. Fox

cessfully survived for millions of years long before we appeared on earth.

Teaching pets "tricks" gives pet owners a sense of pride, accomplishment, control and amusement but a wholly erroneous view of animals' abilities. Many cats are not dependent enough to be motivated to learn often meaningless acts such as rolling over or extending a paw to shake hands. Food reward doesn't work for all pets, either.

Teaching an animal to do tricks for home entertainment or commercial purposes (movies, TV ads and so forth) I find demeaning and exploitative. Even the simplest tricks are too often based upon subordinating the animal. Making animals perform such acts means making them display their submission, which, in gratifying the human ego, says much about our need to feel superior and in control. (However, I am not opposed to basic obedience training, which is useful in helping animals adapt to living with us, and vice versa.) Most extreme examples of animals' being trained to perform unnatural acts are seen at some zoos and at circuses, where lions and tigers jump through hoops of fire and bears balance on stools and rolling balls. These spectacles, which are now banned in a few enlightened European countries, such as Denmark, are a sad reflection of our need to demonstrate our superiority over animals and willpower to control and subordinate them. However, some tricks can be taught to animals that provide us with the opportunity to investigate their intellectual abilities of reason and insight.

There are a number of tests you can use to evaluate how well your cat can think and reason. For example, set up a barrier with sufficient space beneath for your cat to extend its paw. Place some meat in a pie tin and attach a string with a piece of wood tied on the end, placing the wood just in reach of the animal's extended paw. Most cats, if sufficiently hungry, will at once reach out and pull in the piece of wood to get the meat. This is termed insightful behavior.

Insight and foresight and the ability to extrapolate are seen frequently when cats are playing, hiding and ambushing each other, just as when they are hunting prey and anticipating the prey's actions.

Many cats demonstrate well-developed discrimination and anticipation, basic ingredients of thinking and reasoning, as when they become excited on hearing the voice of a family member. In their ability to discriminate different odors, our pets are also far more intelligent than we.

Do animals sit and wonder or worry about things as we do? That

Trained animal acts demonstrate man's desire to exercise control over the minds and actions of other species. In some countries of the world, these displays are now prohibited by law.

Photo courtesy M. W. Fox

is difficult to know, since they can't speak, but it's quite clear when a cat is anxious or apprehensive. Cats also dream, twitching and even mewing in their sleep. Clearly they must be able to imagine things, another basic component of the ability to think and to recall and to recognize. There is another belief promulgated by some eminent scientists that, since animals have no language, no word-symbols, they cannot think. But this is clearly erroneous: there is a whole realm of preverbal memory and mental activity that animals share with us and that our verbal abilities can either repress or enhance, as most psychoanalysts and therapists will attest.

The following well-documented examples from my records will help demonstrate animals' abilities to think, at times very logically, and to use foresight (that is, set up goals) and hindsight (that is, learn from experience) in their carefully thought-out behavior.

1. *Behavioral control and manipulation (of owner):* Many cats meow and solicit incessantly until they are picked up and petted or given a treat. They are clearly filled with expectation and the knowledge that their owners will respond; otherwise they wouldn't persist in doing what they do.

2. *Observational learning:* Many cats have taught themselves, after observing their owners, to press doorbells, flush toilets and switch lamps on by pulling the chain-switch. I have no records of tool use per se (which is a clear demonstration of forethought) in cats, although sea otters will use stones to crack open abalone shells, and even "small-brained" herons will drop a feather in the water as a lure to catch fish. Some forms of intelligent behavior go beyond simple observational learning and mimicry. One correspondent described how her untrained cat presses its paw on the bar of the electric can opener in the kitchen to produce a loud noise that signals it wants to be fed. Has the cat conditioned its owner, too?

3. *Symbolic behavior:* Many cats express their wants and intentions with some nonverbal symbol rather than relying exclusively on body gestures or vocalizations; for example, by presenting their owners with a ball or other toy when they want to play.

4. *Emotional disturbance:* Symptoms are often comparable to those seen in humans, including separation anxiety, grief, depression and anorexia nervosa, fear, jealousy, guilt, and

This fastidious cat is demonstrating an instance of *observational learning*. It is not uncommon for cats to adopt this form of using the amenities around them.

Photo courtesy M. W. Fox

The household cat's ability to manipulate humans has been perfected over the centuries. In most cases, when a cat wants to be picked up and petted, it is going to be picked up and petted!

Photo courtesy HSUS

9

psychosomatic disorders such as diarrhea, pruritus, epilepsy and neurotic fur-pulling. All indicate subjective states of mental activity in many ways comparable to our own.*

5. *Insightful behavior (and reasoning):* Wildcats have been seen tapping the water surface at the edge of a stream with one paw, copying the movements of a fly to catch fish. A cat waiting in ambush around the corner for a playmate to appear is clearly showing insightful behavior and reasoned expectation.

6. *Imagination:* Evidence of imagination (or what ethologists term a mental "search image") can be inferred from the movements and sounds cats make when they are dreaming. Likewise, a cat at play may pounce upon and chase nonexistent "prey." Such vacuum activity or hallucinatory play indicates imaginative ability.

7. *Sense of humor:* Seemingly aggressive behavior that occurs in playful teasing and mock attack necessitates a degree of awareness that such behavior is not serious; otherwise, an aggressive or defensive response would be elicited. The cognitive shift in correctly interpreting such behavior as nonserious implies a sense of humor, an intrinsic feature of social play in cats and other mammals.

8. *"Minding" behavior:* This entails ability to put oneself in another's position empathetically. It is most often demonstrated when one cat is sick and another takes care of it, sleeping close by and grooming it frequently. Cat owners also describe their pets' altruistic behavior when they themselves are sick and their feline companions become more concerned and attentive. Such altruism in animals is surely a sign of highly developed ability to "tune into" and understand others, an aspect of intelligence and emotional awareness that is not exclusively human.

9. *Evidence for a sense of self?* Cats have a memory and thus a reference point to self (an observing ego?) in time. They manifest territoriality and thus a sense of self in space. In their sociability and social hierarchies they must also have a sense of self in relation to others more dominant or subordinate.

* For details, see M. W. Fox (editor), *Abnormal Behavior in Animals* (Philadelphia: W. B. Saunders, 1968). Also *Understanding Your Cat* and *Understanding Your Dog* (New York: St. Martin's Press, 1991).

When pet cats engage in mock battles they are often demonstrating the existence of imagination and even a sense of humor. *Photo courtesy M. W. Fox*

Minding behavior points to a sense of altruism in cats as well as the ability to express social responses within the group. *Photo courtesy M. W. Fox*

Many animals seem aware of death, and though they are less preoccupied by the fear of it than humans are, they must sense it and experience grief—emotional separation and loss.

Some believe that such awareness brings with it the glimmerings of religious inclinations and spiritual awakening. Have animals a conception of divinity, or are we the lesser gods of cats and other creatures under our dominion? I believe this is so and that we must act as gods—with compassion, respect and understanding toward our "sister and brother" animals, as St. Francis of Assisi used to say.

Plato and Pythagoras, like Albert Schweitzer and Mahatma Gandhi, would have cautioned that if we abuse our power of dominion and demean or exploit animals unethically, or flatly deny our kinship, then we also deny our kinship with God and assume dominion over the Creator. This is the sin of *hubris*.

But *do* animals have any religious sense? Charles Darwin emphasized that the mental differences between man and the higher animals was not one of kind but of degree, and that intelligence, language and religion were not uniquely human attributes. He recognized the body postures and sounds that animals made as their language, observed that dogs have a conscience and saw their moral sense and deep affection and loyalty for their owners as approaching religious devotion. It is wrong to discuss such animal attributes as being fictitious, sentimental projections, so-called anthropomorphic thinking, of giving animals attributes only humans possess.

Professor Donald R. Griffin*, noted ethologist at Rockefeller University, writes:

> It is actually no more anthropomorphic, strictly speaking, to postulate mental experiences in another species than to compare its bony structure, nervous system, or antibodies with our own. Subjective qualities have remained largely untouched by the Darwinian revolution, primarily for lack of effective methods for detecting them reliably in other species, let alone analyzing them by scientific methods. The prevailing view implies that only our species can have any sort of conscious awareness or that, should animals have mental experiences, they must be identical with ours, since there can be no other kind. It is this conceit which is truly anthropomorphic, because it assumes a species monopoly of an important quality. The attitude resembles, in many ways, the pre-Copernican certainty that the earth must lie at the center of the Universe.

* *Animal Awareness* (New York: Rockefeller University Press, 1985).

THE PROS AND CONS OF ANTHROPOMORPHIZING

Many people take it for granted that we can communicate with our companion animals—horses, cats, dogs and parakeets—and that to a degree we can understand them and they us. They can anticipate our actions, often sense our mood, greet us, obey and disobey and even play with us, and learn from us also. These commonplace animal-human relations are too easily dismissed as being attributable to anthropomorphic thinking and to domestication's influence upon these animals. But when fear reactions do not intervene, humans can have very similar relationships and experiences with nondomesticated animals: with orphan human-raised squirrels, skunks, parrots, wolves, tigers, deer and gorillas. Some wild animals (when there is no fear) will interact socially with us. The playfulness and altruism of dolphins have been documented, as have the sociability and trust of whales and elephants.

Our companion animals—emotionally, socially and cognitively—are indeed more similar to us than they are different. I believe that most people do not "humanize" their animal companions. Rather, the animals humanize us by responding to us affectionately, obediently, trustingly, sensitively and intelligently. And when we respond to them in like fashion, we are not being anthropomorphic, we are being humane. This kind of reciprocal relationship between human and feline is expressed lyrically and perceptively by May Sarton in *The Fur Person:*

> [A] cat whom human beings love in the right way, allowing him to keep his dignity, his reserve and his freedom . . . who has come to love one or, in very exceptional cases, two human beings, and who has decided to stay with them as long as he lives. This can only happen if the human being has imagined part of himself into a cat . . . just as the cat has imagined part of himself into a human being. It is a mutual exchange.

Biologically, animals are our distant kin, and when we relate to them with feeling, understanding and respect, we affirm this kinship, which is not an anthropomorphic notion. It is a scientifically verified fact. Konrad Lorenz, Nobel laureate and one of the founders of ethology, the scientific study of behavior, once stated, "The similarity [between animals and man] is not only functional but historical, and it would be an actual fallacy not to humanize [i.e., anthropomorphize] to some degree." Animals, like us, possess varying degrees of sapience and sentience. This is our biological and spiritual kinship with them.

Without an understanding of feline social signals and body language, it is easy to think that cats are aloof or even unfriendly. Actually, a cat's interactive language is very subtle and easy to misinterpret. What, for example, can be determined from the facial expression of this cat?

Photo courtesy HSUS

2

Know How Your Cat Communicates

CATS ARE SAID TO BE INSCRUTABLE, uncommunicative and aloof. Indeed, to some people they are, because their behavior is extremely subtle and is often either misread or simply overlooked. Certainly, cats are not so gregarious as dogs: while a friendly dog may jump onto your lap, a cat is being extremely sociable when it decides to get up and sit two or three feet away from you. Sometimes when you look at a cat, it will close its eyes and seem to ignore you, as though confirming the myth that cats are distant and aloof. What the cat is actually doing is relaxing in your presence. A tense cat would certainly not close its eyes and would stare and perhaps back away when you look at it. But a friendly cat, one that feels comfortable in your presence, will often close its eyes and relax as a very clear social signal that it accepts you and feels good in your presence.

First of all, let's take a good look at the outside of the cat before discussing its inner psychology and overt behavior. The cat's whiskers are thought to be touch receptors that help the animal negotiate its way through narrow holes. More likely the whiskers play a very important role in protecting the eyes when the animal is going through dense cover; as soon as the whiskers are touched the eyelids reflexively close and therefore afford immediate protection from injury. The whiskers are also extremely pressure sensitive, so that the animal can monitor the direction of the wind constantly—which is essential when hunting and tracking. Cats and dogs are generally thought to be color-blind, but, in fact, cats are sensitive to red and green. Their eyes lack the extreme color sensitivity that we have, but they compensate for this lack by being acutely sensitive to movement. The cat's ears, like the dog's, are exquisitely sensitive to sounds, and both cat and dog can hear high frequency sounds that are inaudible to the human ear.

Both cats and dogs have a second organ of scent located behind the upper front teeth. This is called the organ of Jacobson, or vomeronasal organ, which connects with brain structures associated with sexual behavior, aggression and territorial behavior, and may well be responsible for the profound effect that certain odors can have on our pets' behavior.

The cat has a number of facial expressions that display its emotions and its intentions. A dominant cat keeps a cool front when facing up to a rival, showing little or no expression: a face of passive indifference, with ears upright, whiskers straight out and eyes looking directly and fearlessly ahead. A contented cat may even close its eyes, and a fearful one will flatten its ears back and hiss, and has wide-open eyes with clearly enlarged pupils. A more ambivalent cat, torn between flight and attack, will twist its ears sideways and give an almost cross-eyed stare—and its pupils will be moderately enlarged.

Another facial expression that cats often show is the open-mouth gape, or Flehmen reaction, as when they have sniffed some novel odor or the urine of another cat. The mouth is open and the eyes half-closed, as though the animal is momentarily "stoned."

In addition to its facial expressions, the domestic cat has various body postures that also communicate its intentions and emotional states, and one can see the shift in mood as the animal varies the position of its head, the angle of its body and so on. A dominant cat stands upright, with head held low, as though ready to attack, when facing a rival. A subordinate cat will assume a defensive crouch. Cats do not roll over onto one side as dogs do to display submission. Back arching in the cat is one way of increasing the apparent size of the animal: a kind of illusory trick. It is a defensive posture and is an indicator of intense fear as in the archetypical Halloween cat.

Besides these body postural displays, cats perform various acts to communicate their intentions and emotional state. One very obvious

A cat's facial expression will vary considerably based on changing stimuli (modified after Leyhausen, 1973). The most obvious changes can be seen in ear position and dilation of pupils as shown in these drawings: a) normally alert; b) aware of offensive threat; c) showing increased fear; d) ambivalent offensive-defensive expression; e & f) shift to increasing defensive threat.
Courtesy M. W. Fox from Understanding Your Cat, *Coward, McCann & Geoghegan, New York.*

one is rolling or flopping over onto one side. It is usually an invitation to play. Repeated rolling and calling is seen in female cats during courtship and when they are in heat. An owner may think that his cat is having a seizure or has a brain tumor, especially if he has a Siamese cat who is rolling and calling continuously. Rolling over onto one side is a common prelude to play in kittens. Cats engage in a variety of play action, including solitary play, such as playing with a leaf, a ball of wool or their own tails, and this to me indicates that they do have imagination. Sometimes they will chase after some imaginary, nonexistent object between their front paws, and at other times leap up and bat at a fleck or mark on the wall as though it were a fly.

It is important for owners to realize that cats enjoy playing. Animals and people who play together stay together. Play is not only good physical exercise, it has a bonding function, too. Cats enjoy games, including play fighting, wrestling, chasing, stalking and games of ambush and scare-me. The scare-me game is best exemplified by a cat walking up to a wall mirror, back arching and then running away as though terrified. A person can play this kind of game, scaring the daylights out of the cat and playing hide-and-seek and catch-me-if-you-can. Because of the cat's need for and enjoyment of social play, I do advise people to keep two cats rather than one, especially if they leave the animal alone all day when they go to work.

In relation to this issue, the cat is, I believe, going to be a major companion animal of the future. It does not need to be housebroken, will readily use a litter tray, doesn't need to go outdoors for exercise or to urinate or defecate as a dog does, and adapts extremely well indoors. However, it does need companionship, and there is nothing better to keep one cat healthy than another one to play with, to sleep with and to engage in reciprocal social grooming. Unlike cats, dogs, when left alone, will often become house wreckers and develop various emotional behavioral problems.

Some cats will develop behavior problems as a consequence of boredom, one being excessive licking or self-sucking, another being wool chewing or chewing various other materials in the house. This chewing vice is particularly common in Siamese cats and might have a heritable basis to it. Boredom, I believe, is a major contributing factor, which cats communicate indirectly via their "neurotic" behavior.

It is not generally recognized that cats use their tails as a very important means of communication. A cat that is excited or aggressively aroused will lash its tail. Some cats, when talked to, will move

The vertical tail display is a positive signal. It can mean "follow me," indicating friendly intentions. Kittens following their mother will instinctively carry their tails in this manner.

Photo courtesy M. W. Fox

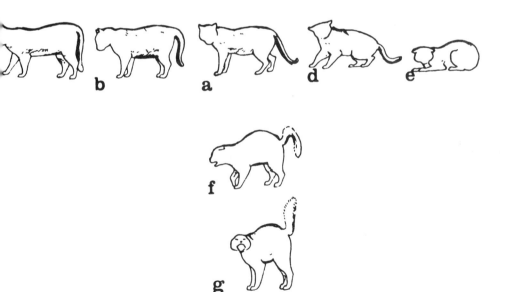

Body postures responding to different stimuli (modified after Leyhausen, 1973): a-c) increasing offensive display; a-e) increasing passive submission. Cat may roll over into defensive posture (e). a-g) increasing defensive threat display.

Courtesy M. W. Fox from Understanding Your Cat, *Coward, McCann & Geoghegan, New York.*

the tip of their tail. Mother cats will move their tail as a play stimulus for their kittens. Tail position plays an important communicatory role in cats. The vertical tail is a social signal in kittens, indicating friendly intentions or soliciting attention. It is also used by the cat to attract the owner's attention, especially first thing in the morning when it wants to be fed, and many tails have been lost in refrigerator doors as a consequence of this display! There is another form of communication in cats that can be a source of annoyance, if not actual hazard, first thing in the morning when a cat is soliciting food. The animal will suddenly walk in front of you, directly across your path, and this is its way of telling you to follow it. A vertical tail could then be used as a signal to elicit the following response.

When a cat is being petted, it will also raise its tail in a vertical position. This display could be related to the kitten being cleaned at the hind end by the mother. In other words, this may be a socioinfantile display related to early interaction between a kitten and its mother, later persisting into maturity as a social, affectionate and solicitous signal. Adult cats may display other infantile behaviors when being petted, including drooling, pushing with its head, as though trying to nurse, and alternate kneading with the front paws. This is a perfectly natural, affectionate and solicitous "regression" to kittenish behavior, another example of socioinfantile behavior that some owners mistake as neurotic or perverted.

Other tail displays in the cat, in addition to the vertical friendly tail, include the fluffed-out vertical tail of intense fear and defensive aggression. An alert, excited cat will show a tail that is stiffly erect, with the hair fluffed out, or an inverted U-arched tail during playful situations. This, from my research in foxes, seems to be a very clear play signal to solicit chasing, because in foxes and in cats the tail is a very potent social tool signaling playful intentions. In foxes the tail might actually be presented for another one to grab, and then a chase or a playful fighting ensues.

One cat will sometimes mount another and grasp the scruff of the neck to assert dominance. This is often misinterpreted as sexual behavior. Seizing a cat by the scruff has an immobilizing effect on the animal and is a very effective way of asserting control over a cat that is difficult to handle. This behavioral response to having the neck seized may be related to the mother's seizing and carrying a kitten by the scruff when it is very young. This socioinfantile reaction persists into maturity as a sociosexual response, as when the female cat re-

In typical "chasing play," the invitation to chase is signaled by a "U"-shaped tail position. The vertical tail position indicates the excited state of the "pursuer" in this case.

Photo courtesy M. W. Fox

When kittens interact, roles can change with blinding speed. The tabby kitten doing the chasing has turned its tail into a "U" and it is now the subject of the chase.

Photo courtesy M. W. Fox

mains passive when the male seizes her by the scruff prior to mating. When one male does it to another male, this has been termed psychological "castration" and is a sure sign of dominance.

When two cats interact, eye contact plays a very important role in regulating social distance. The dominant cat will give a direct stare to a more subordinate one. When the dominant cat breaks eye contact, the subordinate cat is then allowed to move away. The subordinate cat usually moves slowly because if it were to run, it might elicit the chase response and be attacked by the dominant individual.

During social encounters between cats, nose touching frequently occurs. It is possible that the cat is checking the familiarity of the odor of the other cat: it is a fact that cats who know each other well will fight when one individual in the group has been out and smells of another strange cat, or has been to the veterinary hospital and comes home smelling quite different. Sometimes when two cats are facing up to each other, there is a good degree of ambivalence before they work out which is the dominant of the two. When a cat is in a state of ambivalence or conflict, it will engage in so-called displacement behaviors. A displacement behavior is a behavior pattern that is not strictly appropriate for the context, such as grooming in the middle of an aggressive confrontation. A cat in an aggressive or tense social situation may suddenly start grooming itself, and such displacement behavior might serve to "dearouse" or relax the more ambivalent animal.

It is often said that a person can tell when a storm is coming because his cat will engage in bouts of rather intense displacement grooming. This could, in fact, mean that the cat is feeling tense and anxious because of the change in barometric pressure or ionic charge of the air prior to a storm. Cats, when disciplined—say, being told to get off the table—often groom themselves briefly after being reprimanded—a sign of embarrassment perhaps.

Cats' tails also have an additional function of scent marking. They will brush their tails across furniture, for example. The tail has diffuse scent glands under the skin, and it is quite likely that such marking makes the animal literally feel at home by imparting a familiar odor to its surroundings: not so much a territorial mark but a marking of home base in order to feel secure. Cats will also mark furniture and their companions, including people and other cats, with their temporal glands, scent glands located on the temples in front of the ears. They also have a gland under the chin and on each side of the lips for social and home-marking. In the cat world those who smell the same are

Nose touching is a common social device in cats. It is used to identify individuals and to establish dominance. After being checked out by the dominant Abyssinian at the right, this kitten is advised, in no uncertain terms, of its place in the social order.   *Photo courtesy M. W. Fox*

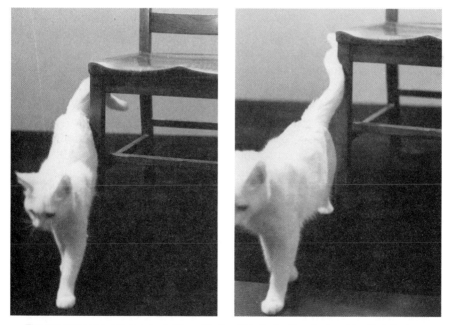

The cat shown here is scent marking objects within its territory by brushing its tail where it wants to leave its mark. Glands located just under the skin spread the scent that sends important messages to cats living in the same area.   *Photo courtesy M. W. Fox*

23

regarded as friends, and so it is logical for cats to engage in a good deal of reciprocal marking.

Cats have another means of scent-mark communication, and that is spraying objects with their urine. Such behavior is more common in entire male cats than in female cats but also occurs in females and in castrated males when they are motivated to defend their territory. The presence outdoors of a rival cat that is spraying, or any change in the social relationship within the cat's home, can make it feel more insecure and therefore more likely to mark or spray within the house.

As for a cat's refusing to use the litter tray and defecating in various places in the house, this again can mean that the animal is emotionally upset because of some change in the social relationship within the family, such as the birth of a child or a change in the social hierarchy if there is more than one other cat in the home. Wildcats such as bobcats will fastidiously bury their feces close to home base but leave their feces unburied at the edge of their territory as territorial marks. Hence the interpretation that if a cat refused to use its litter box and leaves its feces uncovered in various parts of the house, it is feeling insecure and has a greater need to mark its territory. There are many emotional reasons why cats become unhousebroken, and good behavioral counseling can save many cats from euthanasia. Of course, we should be alert to the fact that constipation and pain in passing hard stools can set up a condition aversion, and cats will then refuse to use the litter trays. Also, cats with cystitis may urinate in various places in the house, sometimes even at the owner's feet, as though to communicate that they are sick. The sheer pain of passing urine when the animal has cystitis can set up an association that makes the litter tray seem the source of pain, and it is then not used.

Another point of general interest related to cat behavior is the cat's tendency to paw and scrape around its food, as though to bury it. A cat will naturally bury food in the wild and sometimes return to eat it. At other times the reaction might well be one of disgust because of the smell of the food, and the cat is literally attempting to bury it and cover up the smell!

Cats frequently suffer from fur balls, which can cause intestinal blockage, but more often a healthy cat will engage in frequent bouts of retching and throw up the fur balls. It is perfectly normal for cats to swallow fur as a result of grooming with their raspy tongues, but possibly because of the diet that we give them, there is not sufficient roughage and bones and other materials to bind up the fur so that it can be passed out in the feces normally. Hence the fur builds up in the stomach, and the cat should be able to vomit this up. A cat that is losing condition may well be suffering from fur balls and should be given mineral oil or a mild laxative.

Cats do enjoy eating fresh grass, which may be a stimulant or a general tonic. It may also help in the passage of fur balls. Owners with confined cats should be encouraged to grow a little grass in a window-sill pot.

One of the ethical questions we should raise is how far we should go in making our pets adapt to our life-styles. Declawing cats, for example, is a particularly touchy subject. I do not believe that animals should be mutilated unnecessarily, particularly the ear-cropping of dogs. Cats enjoy scratching, which may be a social display and a way of marking territory. They should be encouraged to use various scratching posts that can be "spiced" with a little catnip to make them more attractive.

Cats are indeed fascinating, intelligent and affectionate animals. Possibly because of the subtlety of their behavior they have been misunderstood by man and even regarded as unresponsive, distant and aloof by many people. But they do make extremely satisfying companion animals, and a greater understanding and appreciation of their behavior and emotional needs will go a long way toward improving their relationships with their owners and the treatment they receive. And the better we can communicate with them, once we understand how they communicate, the more their potentials will blossom and their "catness" be fulfilled.

It is interesting to note similar behaviors in species from different genera. The back-arching threat display shown by this grey fox is the same as for the margay, a small species of wild felid, and the Abyssinian shown here. *Photo courtesy M. W. Fox*

3

How to "Decode" Cat Behavior

$\mathbf{T}$HERE ARE OTHER ASPECTS of cat behavior that can be difficult to "decode." What is involved is a careful observation of the animal's behavior and development in all kinds of contexts, to enable one to ascertain what motivates ("turns on") certain behavior and what the outcome is for the cat in behaving in a certain way.

But in order to be able to decode your cat's behavior more fully, such as understanding why cats rub their heads against you or spray the house, I have found my studies of wild carnivores—foxes, coyotes, jackals and wolves—and other research on wildcats, lions, bears and such most helpful. These species are distant cousins of cats, and an understanding of their behavior allows an evolutionary, social and ecological perspective, which is essential because some of the behavior of our feline companions is influenced by domestication. (Domestication includes both genetic, developmental, social and environmental influences on an animal's behavior.)

Cats often lick their owners' hands and seem to enjoy it. Again, some pet owners interpret this behavior wrongly, believing the animal likes salt (from our sweat) and so licks our hands and faces. It is not difficult to decode this behavior when we see how cats interact socially with their own species. They are not licking salt off each other, they are being affectionate and attentive, bodily care being something that can be shared in a bond of empathy when they engage in social grooming.

Some behavior is especially difficult to decode because the signals are ambiguous and overt in different contexts.

For example, a cat will roll over onto its back and twist around because it is soliciting play. Female cats also behave this way when they are in heat and don't want to play. When in heat, the rolling is more intense, and the cat may give long, mournful calls as well, and lick its hind end frequently.

Some behaviors are difficult to decode for other reasons. One is that we don't know why the animal is behaving in a certain way. A cat may pace the house giving long, plaintive calls because it is in heat and is calling for a mate. It may also be in pain and suffering from cancer or infectious peritonitis. To be sure what the pet's behavior means, it is a good idea to check with your veterinarian. Cats in pain often seem depressed or afraid, while some dogs act submissively as though they are being reprimanded.

Misunderstandings may arise when the animal's behavior occurs out of its natural context; examining the behavior in its natural context can help us understand some bizarre and often irritating behavior of our pets.

One example is a cat's bringing dead or half-live birds and mice home. These would be for the cat's kittens in the wild, for food, and to teach them how to kill. A humanized cat redirects this natural behavior toward its human family, who should accept such gifts with understanding and gratitude and not encourage their cats to roam free and needlessly kill or injure wildlife.

Adult cats will often drool and "make dough" by kneading with their forefeet against their owners when they are being petted. This is a regressive, redirected behavior: the cat is behaving like a kitten and redirecting its infantile nursing behavior toward its mother-substitute human companion. We should accept such behavior as a sign of absolute trust and affectionate dependence.

Cats and other animals, like many humans, are affected by the full moon, during which time there is a greater negative ionization of the atmosphere that makes many of us more active or "charged." Microscopic iron particles in the heads of cats, bees, pigeons and even humans account for the uncanny behavior of navigation, of lost pets being able to find their way home. They have compasses in their heads!

The female Siamese shown here is in season and is rolling and twisting to solicit the attention of her suitor. Cats often solicit play in the same way but with must less intensity than is characteristic for the in-season queen. *Photo courtesy M. W. Fox*

Long after weaning, some young cats will indulge in a form of nursing behavior. This tabby is nursing its owner's arm—a completely natural action. *Photo courtesy M. W. Fox*

One behavior that science cannot explain but that we can intuitively appreciate is the psychic behavior of pets, like that of the cat who suddenly began to meow loudly, according to its master, at the precise time when its companion dog was with its mistress at the veterinarian's being euthanized.

Many new cat owners have expressed concern that reflects their misunderstanding over how their cats sometimes display affection. Cats do get carried away when they are being petted, and a very common reaction is to begin to drool and knead with the front paws, as described previously. Other cats, however, will suddenly chomp on their owner's hand or arm, and sometimes this can be quite painful. This should not be interpreted as aggressive behavior unless, of course, the cat bites hard and then rakes you with its claws. Some cats will do this because they do not like to be petted too much or for too long. It is advisable in such instances to pet them as little as they like. The bona fide love-bite tends to be somewhat inhibited, and the cat literally holds on for a few seconds. It may even attempt to straddle your arm as though it were mounting a mate. Such explicit sexual behavior should not be seen as a perversion, but a natural reaction that is to be expected from an animal that has a close, emotional bond with a companion who, because of the situation of domestication, is not another cat but a human being.

We, as observers and interactors with our pets, also affect their behavior. The observer effect, in which your presence per se affects the animal's behavior, can be seen when your cat looks at you and starts rubbing its head, lips and tail on the furniture in front of you. Your presence is making your cat give you a social display of affection. This is the cat's way of telling you that if you were to come closer, it would rub against you and anoint you with the scent glands on its tail, forehead and lips.

One example of complex conditioned behavior concerns a cat who would put its toys into its empty food bowl whenever it was hungry and its owner was late coming home from work. Like using tokens or money (which monkeys have been trained to use to obtain various rewards), this cat most likely associated putting its token toys in the food bowl with receiving food as a reward. The cat learned that its owner would exchange the toys for food, the owner's behavior thus rewarding and reinforcing this cat's unusual behavior.

Another cat developed the habit of pressing on the lever of an electric can opener when it was hungry. Such behavior demonstrates

Rubbing against a person is a cat's way of showing affection for a human. It also allows it the opportunity to scent mark that person. *Photo courtesy M. W. Fox*

31

cats' ability to train their owners, communicate their wants and learn through observation. Other cats have learned to use and flush toilets, open doors, press doorbells to have the door opened for them, switch on lights, radios, televisions, and even knock telephones off the receiver when they ring, and mew into the mouthpiece. The interactions between observant cats, their owners and these various devices can be construed as evidence of a highly developed mimetic capacity in cats, which, because cats are so dexterous, can greatly surprise their owners. Naturalist Hope Ryden has seen bobcats at the edge of a stream patting the surface of the water with a paw, as though mimicking a fly's struggles, in order to catch fish, at which they are quite proficient.

A visitor in the house who stays for several days can so upset cats that they become unhousebroken. Decoding this behavior as a sign of emotional arousal and insecurity, because the cat's territory has been intruded upon, is facilitated by an understanding of wildcat behavior. As described previously, around home base bobcats carefully bury their excrement, but at the edge of their territories, they leave all their evacuations exposed so that other cats will be able to easily detect their territorial boundaries. The poor, insecure house cat, who is often punished and misunderstood for such wanton behavior, is simply marking its territory in order to feel more secure and perhaps drive away the intruder. Sometimes cats outdoors trigger this behavior as they invade the resident cat's porch or yard. Often a newborn baby triggers this territorial marking, or a new roommate or spouse, or even sexual maturity—especially in young males—hence the sensibility of neutering them before maturity.

CATS AFFECTED BY OWNERS' STRESS

A very common reaction of cats to stress is intense licking and even pulling out of the fur along the stomach and the back of the legs. This can become so severe that the cat becomes quite bald. Veterinarians recognize this as an indicator of feline frustration, and it is often linked directly to owner anxiety. According to veterinarian Dr. Robert Whitney, this problem is on the increase, especially in the Midwest where he practices. This type of feline stress was rarely heard of fifteen or twenty years ago, but today he is treating eight to ten cases every month. He is not alone in feeling that one of the common stress factors may well be the economy, and that owners who worry about their jobs and maintaining their life-styles may unconsciously pass their anxiety on to their cats.

Because of the close emotional bond between the cat or dog and its human companion, it is not surprising that the emotional state of the latter can have a profound effect on the emotional state and overall well-being of animal companions. I have termed this phenomenon "sympathetic resonance," and it is something that we should all be aware of, since it is not widely appreciated to what extent our emotional state can harm or benefit our animals.

Knowing the natural behavior of animals outside of their domesticated contexts and being able to stand objectively outside of my relationship with an animal have certainly facilitated my ability to decode animals' behavior. And we all have the ability to empathize, to put ourselves in the animal's place, which is as much a prerequisite for good relationships with each other and with animals as it is for the scientific study of behavior and animal care in zoos, laboratories and "factory" farms. When we realize that the economy of animals is such that they do not behave without good reason and that, unlike us, they are almost incapable of concealing their emotions, intentions and desires, we may acquire sufficient respect to really pay attention to what they are trying to tell us and what they are communicating, so often unintentionally, with the transparent honesty of little children.

We have learned that animal sounds are more than noise. Animals vocalize in specific patterns for a wide variety of purposes. If you listen to your pet cat or to a queen with her kittens, it is very obvious that animal sounds are in every sense a form of communication.

Photo courtesy HSUS

4

"Felinese" Sounds
That Make Sense

CATS MAKE A VARIETY OF SOUNDS, as do most other animals, that most people think are a lot of noise. But when we stop thinking that animals lack intelligence and cannot really communicate their wants, feelings and intentions, we begin to listen to them. And when we really listen, we begin to hear what is primarily an emotional language of sound. This is interspersed with specific sounds that have symbolic or objective meaning, as the alarm bark of some species of monkey to a ground predator such as a human or jaguar, which is quite distinct from the call given when there is an aerial predator such as a monkey-eagle. With one bark they know they must run up the tree, and with the other they know they must run down.

Now some academicians, who might profit by stepping out of their intellects and into the real world of the emotions for a moment, insist that animals have no language as we have. Since there is no grammatical structure in animals' sounds, it is claimed that they have no language or linguistic ability per se. But one definition of language, according to *Webster's Dictionary,* is "any means, vocal or other, of expressing or communicating feeling or thought." Another is "the inarticulate sounds by which animals express their feelings." (Mechanistic scientists examining animal behavior won't like that!) That animal sounds, to our unattuned ears, seem inarticulate does not mean they are unintelligible (or "dumb"), but rather they are not articulated together into segmental syntax and grammar, as in our word-sounds. In our language, we can give a stream of multiple images, ideas, desires and feelings, while in animals' language, the stream is narrower but no less meaningful, and a spectrum of feelings, desires, images and intentions are expressed.

I used to think that cats meow, purr, hiss and scream instinctively, not intentionally to communicate anything consciously. They seemed to make these noises automatically, just because they felt a certain way in a particular context. However, there is conscious intent, because these vocal sounds are made only in certain contexts and not in others. When I come home, my cat Sam emits a loud meow-purr of greeting, then a more plaintive meow. Distress or protest at being left alone all day? Then he puts his tail up and makes a beeline for the refrigerator. He's saying "meow" (me), "distress" (hungry). So I feed him. Sam is demonstrating anticipation of some future event in his plaintive meow and reveals his intentionality in all that he says and does.

Animals are as aware of context as we are. Perhaps because of their relatively limited language compared to ours, they must be acutely aware of context in order to avoid ambiguity of signals. For example, a low growl or hiss to other cats at feeding time will repel them, but a low growl when out in the open may well send them running for cover. Both hiss-growls are warning signals. They may not be identical: analysis of these sounds with a spectrograph may identify subtle differences not intelligible to our ears. The mate-identification calls of sea gulls, for example, all of which sound the same to our ears, have subtle differences in frequency (detectable on an oscilloscope or sonograph) that enable look-alike sea gulls to recognize each other.

THE MEANING OF SOUNDS THAT CATS MAKE

Cats make a variety of sounds, some "pure," like a meow, others more complex and mixed, like a purr-meow. All these sounds have various meanings that express the animal's emotional state or intentions. There's some ambiguity, too, because they often give similar sounds in different contexts, like a cat's purring for attention or out of sheer enjoyment. So it's important for us to not only listen to what the

Sometimes a good bluff is just as effective as actually attacking another cat. Here a kitten's defensive-aggressive display actually turns away an adult cat. *Photo courtesy M. W. Fox*

Within a group, cats will often establish social standings with the use of vocalization along with physical signals.
Photo courtesy Myanmar Birmans/David Hovell

cat is saying but also to put ourselves in its place to try to ascertain what it wants and what it is feeling.

Common Cat Sounds

Mew of kittens—distress, cold, hunger.

Purr—contentment, relaxation; *loud purr*—soliciting attention or close contact.

Cough-bark—a rare sound, signaling alarm.

Hiss—threat, fear.

Meow—distress, attention-seeking, food-soliciting (often linked with a purr), calling to kittens.

Chatter or twitter—excitement, as when looking at birds.

Snort-growl—threat, challenge.

Yowl-scream—threat, defensive or offensive.

Scream—fear, pain.

Meow-yowl—"cat call" in heat, wants out!

Meow-purr (or purr-meow)—attention-seeking.

The "caterwauling" of cats at night is the vocal accompaniment of feline face-offs, often two male rivals, sometimes over a female. You will hear meow-growls and yowls, snort-growls and yowl-screams as one tries to intimidate the other by the intensity and timbre of its voice, by its size and intensity of eye-stare.

This "language" of sounds that I have interpreted functionally can be broken down into these basic categories: to increase or decrease social distance (to threaten or to solicit), to call for contact or to maintain close contact (like cats purring together). Our vocal tone that colors what we are saying is called paralanguage. It expresses, often unconsciously, how we feel—friendly, aggressive, fearful, solicitous or any other emotional state, and essentially mimics the basic emotional sounds animals communicate to each other. Hence they can understand what we say to some extent, because it is *how* we say what we say—the emotionality behind the spoken word—rather than what we say, that has meaning to them. They aren't so dumb after all, are they!

There is no use trying to understand what these signals mean without being able to empathize with the animal, that is, put yourself in its place and, through trial and error, attune your mind to the animal's intentions, expectations and immediate context of the here and now (or just-so-ness of its existential world). Through such learning, we can all come to understand animals better, know what they are

There can be no mistaking this aggressive display. This cat's expression clearly indicates its intentions. The simultaneous vocalization intensifies the animal's threatening stance.

Photo courtesy M. W. Fox

A cat in a defensive posture will use a whole repertoire of sounds along with a number of postures and gestures to communicate its stress exposure.

Photo courtesy M. W. Fox

saying to us and respond appropriately. We have to break free of our prejudices (that animals lack intelligence) and break the habitual mode of looking for meaning within spoken sounds or words only. The meaning of animals' communication is rarely in words, as sounds with specific meaning, but more in the broader realm of context and intentionality.

However, there are some sounds that have specific meaning. A cat's purr is a sign of relaxed contentment. A piercing yelp or scream is a clear signal of pain or acute distress. It is up to us to find what is evoking such distress, the specific sound demanding that we look at the context in which it is being given. And animals give mixed sounds to convey more than one emotion, like my cat Sam's loud meow-purr when he wants to be fed.

It must be confusing for some of our pets to have us sitting quietly, not sleeping (they don't usually bother us then), but thinking, reading or meditating. These states seem alien to them, or is it our lack of communication, our intraversion and withdrawal from the world that they find so disconcerting that they come and sit at our feet or on our laps, or push us with a moist nose? When we become too preoccupied in self-contemplation, thinking to ourselves and withdrawn from the here and now, we lose touch with animals and they with us. It disturbs me that so many people have lost touch with the here and now of the world that they do not believe that animals' sounds make sense. Please, come back and listen! Animals aren't dumb, they have a lot to say.

Sound is a very primal and primitive form of communication, changes in respiration—sudden expiration with alarm (leading to a cough or bark*) or relaxed inspiration and expiration (leading to a contentment groan or purr)—being an evolutionary development of auditory communication that gives information to others about one's internal emotional state.

The more we attune our own ears to these primal signals of animals, the more we can understand their emotions and intentions. Such attunement is indeed *at-onement*.

---

* Cats do sometimes emit an alarm bark.

It is universally accepted that purring is a sign of feline contentment. Being in the arms of an especially favored person is almost sure to elicit the response.

Photo courtesy HSUS/Iyer

41

TALKING ANIMALS

Truth or fiction, it is thought that in ancient times people could talk to animals and understand them because the animals were able to talk themselves. Native peoples all over the world who live close to nature learn as children to recognize the sounds of animals. A shepherd who was not alerted by the alarm call of a deer or the excited cawing of a raven might lose one of his flock to a lion or wolf.

He learned to listen to what the animals were saying around him, because their various sounds had meaning. Without a knowledge of animal talk, the shepherd would not know when a lamb was lost or in distress, or when his flock sensed danger. Likewise hunters listen to the talk of animals around them, knowing that a sudden alarm call could mean that one or more animals had seen him, while their continuous social babble meant that they had not caught sight or scent of him, so he could advance even closer. Some hunters even mimic the courtship and social calls of certain prey species, actually talking to the animals to allay their fear and draw them closer.

It is clearly more truth than fiction that people can talk to animals and that animals talk, too. That these abilities were perhaps overblown into some mysterious power or dismissed as folklore or myth by "civilized" people is understandable. Not living close to animals they would not have the opportunity to learn their rudimentary emotional language and certainly would never comprehend animal talk. By listening to the various sounds that animals make, our forefathers had little difficulty in knowing what animals were feeling and likely to be thinking and doing. To anyone not familiar with animals, this would seem like some incredible power. However, once your ears are attuned to animal talk, along with your emotionality, it is easy to feel and therefore know what they are saying.

Dr. Eugene Morton, with the Smithsonian Institution's National Zoo, has been analyzing animal sounds and finds that when we disregard the words we use, our speech patterns follow the same rules whereby animals talk. For instance, when a person is talking in a friendly voice, as to a pet or baby, the voice pitch increases. Similarly, animals' sounds generally are at a higher pitch and lower intensity when they are being friendly toward each other, tending their young, engaging in courtship and showing submission to a social superior. Whines, whimpers and twittering and purring sounds belong to this category of feeling and intention.

The ability to communicate with animals by understanding their vocalization has always held a somewhat supernatural connotation. However, people who live in close proximity to cats and other species are often able to decipher the sounds made by the creatures around them.

Photo courtesy Bonnie Smith

Human speech has a lower pitch when a person is feeling aggression or is being demanding. Likewise an animal that is asserting its dominance over another or is preparing to attack will emit lower-pitched sounds, from growls to roars. Lizards, birds, dogs and lions all make such sounds, which clearly express how they feel—their motivational state.

Animals can give more complex messages, too, which express more than one emotion and intention. A dog's whining growl expresses both fear and readiness to flee or submit, and aggression and preparedness to attack or defend itself.

High-pitched, high-intensity sounds are part of animal talk associated with alarm, as when a bird or monkey is frightened by a cat or leopard. Humans emit similar sounds when frightened, so it is easy for us to identify and thus empathize with the emotional state of animals who make similar heart- and gut-wrenching sounds.

According to Dr. Morton, the bark that birds, dogs and other animals make is a mixed signal that has high- and low-frequency components. The animal is neither purely aggressive nor fearful but is alert and ready to respond to whatever transpires. For instance, a dog that hears a noise outside its home will bark. Once it sees who or what is making the noise, its bark will change to include growls if it is an intruder, or whines and yelps if it is its owner returning from work. A bark can also be a way of saying "Look at me" or "See what I want," an emphatic way of gaining others' attention, much like a yelp or a whine, though the latter sounds are more solicitous or appeasing. Furthermore, the more excited, alarmed or distressed an animal is, the more frequently it will repeat a particular sound. Dr. Edwin Gould, curator of the Department of Mammalogy at the National Zoo, sees a similarity between such repetition and pause patterns in human speech. A person's emotional state is reflected in the pattern and timing of pauses, one who is excited having a speech pattern with few pauses.

So if we attune our ears and listen with our hearts, it is not difficult to make sense out of the sounds of animals. And it is not some ancient myth or folklore that animals can talk, and that we can talk to them. Our tone and pitch of voice, like theirs, give meaning to what is felt and convey motivational state and intentions. The modern science of animal behavior is confirming what our ancestors knew all along and is also reaffirming our kinship with animals.

It can be said that people who really understand animals—their body language and the ways in which they express their emotions, needs and intentions—are able to talk to them. The importance of such understanding is eloquently expressed by American Indian Chief Dan George:

> If you talk to the animals
>   they will talk to you
>   and you will know each other.
> If you do not talk to them
>   you will not know them.
> And what you do not know
>   you will fear.
> What one fears
>   one destroys.

Hence, it is important for animals that people understand and respect them, and indeed talk to them; otherwise people will continue to be indifferent toward them and even afraid. What one loves and understands one will not mistreat or destroy.

Odors of all kinds constantly impact a cat's sensory perception. Curled in a basket, basking in the sun, or strolling through its territory, a cat is always receiving signals through its amazing sense of smell. This sense directly affects virtually every aspect of a cat's life.

Photo courtesy HSUS/Taylor

5

Understanding Your Cat's Super Sense—Smell

$\mathbf{A}$ CAT'S SENSE OF SMELL is guesstimated as being a hundred thousand times more profound than ours. Relatively speaking, our sense of smell is almost nonexistent, compared to a cat's.

Perhaps cats are more aware than we of what different odors can do to their emotions. Odors can make animals fearful, aggressive, sexually aroused, infertile, and can even kill them. In us at most they can evoke sexual arousal and feelings of "edginess"—fear and readiness to flee or fight (as research with perfume pheromones has shown), or security (like a child's familiar-smelling blanket).

Pheromones are fascinating. These are external odors (chemical molecules) that are absorbed when inhaled, stimulating the central nervous system, just as internal hormones do. Hence they are called external (pher)omones. They can synchronize menstruation in humans and influence estrus or heat in cats and many other mammals.

One of these pheromones, called aldosterone, a breakdown product of testosterone secreted by the testicles, appears in the saliva of boars when they sniff a receptive "porkette." This saliva pheromone puts the female into a trance so the boar can do anything. Perfume manufacturers have put this into both mens' and womens' toiletries. Yet we secrete our own! A strong odor of aldosterone generally attracts or repels women and is ignored by men or evokes "edginess."

Cat lovers and keepers know all about catnip, the herb with the odor that gives cats a natural "high," and know that during this ecstatic state the cat shows some sexual courtship, such as rolling and intense but playful tumbling.

Human society would function very differently if odors so profoundly affected our behavior. I often contemplate the probability that our evolution necessitated our withdrawing from such potent influences upon our brains and behavior, so that we could gain control, objectivity and the ability to rationalize.

A cat will attack its companion cat housemate if the latter comes home smelling of other cats or of the vet's hospital. Some cat owners have been attacked by their pets after putting on expensive perfume containing the (inhumanely derived) anal gland "musk" of civet cats. Strange odors make cats very insecure. Some will even spray in the house or become unhousebroken because they are so disturbed by the odors of cats outside. The pheromones from our armpits, breasts and scalps can turn some cats on to nurse like kittens or mount us sexually—profound, if sometimes inappropriate, contact with reality, but not unnatural, since we, in domesticating them into our lives, have disrupted the appropriateness of their responses.

Consider the confusion to pets when their owners come home smelling of another animal that they have been petting. And reflect also upon the confusion of our pets caused by certain hand and face creams, lotions and sodas that contain a chemical similar to the pheromone in the urine of female dogs or farm animal by-products that smell like food.

Such profound reactions as sexual arousal and aggression are not uncommon. They clearly indicate that animals' consciousness of odors is far more sensitive and responsive than ours. In this realm of reality they are superior to us, or more dependent cognitively and socially upon contact with this dimension than we.

The heightened sensitivity to, awareness and dependence upon the sense of smell in cats may be due to the fact that they have more extensively developed nose-receptors than we have. Cats also have a second organ of scent that we and other primates lack (so that we can

Scent signals can be so strong that they can precipitate an attack on a given cat by others in its group if it bears an unusual odor, such as that of a veterinarian's office.

Photo courtesy M. W. Fox

think reflectively?). This is called the vomeronasal organ, or Jacobson's organ. It is located in the nasal cavity and has two ducts or canals that open behind the upper front teeth, creating a direct conduit to whatever the animal licks and tastes. This organ connects with the amygdala, a brain center associated with sexual, territorial and aggressive social behavior.

Cats have scent glands with which they mark their offspring, close companions, owners and even furniture in the house. These are located on the temples, lips and tail. Hence, cats will rub their heads, mouths and tails against us as a sign of friendliness, no doubt establishing a "smell bond" of oneness. These glands, along with urine spraying, are used by cats to establish and maintain a sense of familiarity, and thus security, with their home base and territory. Hence less secure cats, disturbed by rival cats outdoors or by a new pet, baby or boyfriend indoors, may start spraying to help them feel more secure. So don't punish.

Emotional state can be communicated via the sense of smell in animals. People have often wondered if cats can smell fear. As yet, no research has confirmed this in cats, but the probability that they can is high. Research has shown that rats become extremely agitated when they smell the air blown over them from a separate group of rats that are reacting fearfully to a threatening stimulus. Fearful cats often evacuate their anal glands—a terrible stink, which may be an alarm signal to other animals. In the skunk this has evolved into a weapon of defense.

Our emotions do affect our body chemistry, and I am sure our pets can and do divine our emotional state, if not our personas, from the odors our bodies produce. The sweet smell of puppies, kittens and babies is familiar to humans, sometimes effecting a sense of déjà vu, since smell can evoke deep emotional memories. Some human and animal doctors regard their sense of smell as an extra, almost intuitive, diagnostic tool.

Rats can detect the difference, on the basis of odor, between strange and familiar caretakers and between schizophrenic and non-schizophrenic people. This means that our individual personalities and emotional states are expressed by or influence our body chemistry and thus the chemical odors that we produce. Some of these odors are influenced by the kind of food we and animals eat and the kinds of bacteria that live on our bodies and theirs. Hence, differences in odor can be attributed to such external factors rather than to genetic or racial differences. The "prejudice" that cats often show (as fear or aggression) toward humans of the opposite sex to their owners is more likely due to a different odor related to sex pheromones than to a genuine sexual discrimination. Cats do get attached to the familiar odor of their master or mistress, and if the latter rarely meets others of the opposite sex, pets may react fearfully toward them—aggressive reactions that I interpret as sociorivalry or jealousy. A tomcat may spray the visitor's clothes. By putting their own marks on things unfamiliar or intrusive, animals may feel more secure. They will even mark over another's mark, like the cat who urinates on the exact spot where another cat has sprayed. This could be simply to leave a "calling card" in a communal place, or it could be to obliterate the other's mark. Many cats love to roll in and become intoxicated by the aroma of catnip.

Since animals also enjoy certain odors, we might well wonder just how "high" they can get. Whatever odors do to our animal companions, we should attempt to attune ourselves to this extraordinarily profound sense and not be offended when our cats hiss at us for wearing some new perfume that makes them feel as though the essence of a disembodied civet cat has invaded their territory!

Many kinds of animals show a form of psychic ability. For the cat, however, with its apparently inscrutable manner and aloof response to the world around it, a familiarity with the unexplainable seems fitting. So pervasive was this idea during medieval times that untold numbers of cats were destroyed because it was thought they were connected to the underworld.

Photo courtesy HSUS/Judith Halden

6

Supernature: Cats and Other Animals' Psychic Abilities

THE AMAZING SENSITIVITY, intelligence and awareness of animals is occasionally revealed in so-called psychic phenomena. These psychic or extrasensory feats of some animals and humans are beyond the realm of contemporary objective scientific inquiry, which reflects more a limitation of the scientific mind rather than of science per se.

During the winter of 1961–62, a man named John Gambill died in a hospital in a town in Texas called Paris. It is documented that at the time of his death hundreds of wild geese circled the hospital, calling loudly as though in requiem. One would surely regard this as sheer coincidence, yet this man had established a wild goose sanctuary on his farm. The birds somehow seemed to know. The sanctuary had begun years before, when he nursed an injured wild goose back to health. He released the bird, which returned the following fall with twelve geese, and eventually over three thousand wild geese came to spend the winter safely on his property, which he willed to the state as a permanent bird refuge.

This moving anecdote is one of several that have been well documented in books, newspapers and magazines. I have selected a few that will not only make you wonder about the inner world of animals; they may, as they do for me, jolt you into considering the possibility that animals have a different form or dimension of intelligence, a natural inner wisdom that we humans only rarely experience. Some of the anecdotes I am going to give explain themselves. One should not attempt, at least at this stage of our ignorance, to read more into them than what simply happened. A healthy skepticism and an open mind are all that is needed to enjoy these phenomenal accounts. It is not necessary to seek the ultimate truth or objective, statistical analysis and replication.

In his well-researched book, *The Psychic Power of Animals,* author Bill Schul provides some extraordinary accounts of animals' abilities, some even beyond the grave. A man in Kansas City was awakened by his Spaniel's barking and saved his family from a house going up in flames. The dog had died six months before. A Colorado motorist stopped on a mountain road late at night when he suddenly saw his Collie by the roadside. He found that the road ahead had been washed away and that he would have plunged over a cliff if he had driven on. Yet the Collie had been dead for a year. Schul also reports how bees swarmed over the coffin of their deceased beekeeper in England, then returned to their hives.

Other abilities are no less incredible. A sick mountain goat wandered into a small California town and made her way to the local doctor, who treated her. The following year she returned bringing her sick kid. (I have a report of a bobcat who was rescued from a trap, nursed back to health and returned the following year to show her human friend her litter of cubs.)

Schul also relates how a boy, lost in the wilderness, was kept warm all night by a family of beavers, and how a cow became a "seeing eye" guide for her blind farmer-owner, who never had to train her.

I have a friend whose genealogy goes back to a very old aristocratic family in Ireland. He tells me that whenever one of the family dies, foxes, which are normally quite secretive during the day, come out into the open and congregate around the ancestral home. There are two other old families in Britain that are similarly linked with foxes,

Superstitions about cats persist to this day and will probably always be with us. Happily, they have not acted to dull the relationship that cats and their human friends all around the world find so satisfying. *Photo courtesy Judith Halden*

the wild animals appearing, as though to pay their homage, when a member of the family dies. What connection, what historical kinship exists between these families and the foxes on their land?

Other old European families believe that the sound of an owl hooting close to the house could mean the death of a relative or close friend. Such totemism, or superstitious belief in an animal that appears at certain times, could be a living fragment from the distant past when we all "read" animals and used them as a source of information. American Indians today claim to be able to derive meaning from the flight patterns and direction taken by wild geese and from the behavior of a hawk or eagle above them. Perhaps at one time we had a much closer psychic link with our animal kin. Since we were so close to them, living off the land as they did as gatherer-hunters, it would be only natural that they might become a source of information, indicators of omens, of things to come. To the "primitive" human mind, the rational link between cause and effect could have been vastly different from ours, since to find explanations for certain events, primitive people would have looked for any correlation or coincidence that seemed to fit. The appearance of a raven over a teepee might, by sheer coincidence, have occurred at the time of someone's death. The presence of such a bird subsequently might be interpreted as an omen, and the frequency of occurrence might be greater than what could be attributed to mere chance. Such coincidences make us ponder the nature of reality and wonder about the link between animal and human awareness.

A journalist friend of mine recently had to have his aging dog, a very close companion, put to sleep. My friend, knowing that this was best for his dog, stood quietly as the veterinarian and his assistant gently held the dog and gave it a painless intravenous anesthetic, from which it would never recover. The dog quickly sank into unconsciousness, and then it ceased to breathe. To my friend's surprise, at the moment of his pet's death he saw the following: "The dog's lustrous coat suddenly became dull, and I saw a silvery shape of my dog rise out of the still body, and then it seemed to diffuse into the air." This is the closest thing that he had ever seen of what he interpreted as a visible aspect of the soul.

While it may seem inconceivable that wild animals such as geese, foxes and ravens can know when a particular person has died, we might be less skeptical of pets demonstrating such ability, considering the close emotional ties of some with their owners. There are, in fact,

Cats are celebrated for a great variety of mystifying abilities. One of these is *psi trailing*—finding their way home or locating familiar people in unfamiliar places involving travel over great distances. There are many amazing but true accounts of cats undertaking daunting journeys to be with their people. *Photo courtesy HSUS*

many documented cases of pets acting strangely—becoming panicked, calling in obvious distress or acting depressed when someone close to them dies. Such reactions have occurred when the owner, another pet or family member dies. They have also occurred around the moment of death, when the sensitive pet is at home and the deceased many miles away. Distance does not seem to prevent some animals from knowing.

It has often been noted that when an elderly husband or wife expires, the spouse may die soon afterward. The same is true of pets. I remember, from my veterinary practice in England, one old lady who died soon after her cat passed away. Coincidence perhaps? Why, then, should a healthy Retriever die suddenly after a few days' depression following the death of its beloved companion—a canary? It would seem that in some relationships between both animals and people, the emotional bond is so strong that when it is severed by the death of one, the other not only knows and feels but may also die. I recall a tragic case where a little boy was in a hospital dying of leukemia. One of his pet pigeons somehow found him and perched on the windowsill of his hospital room. He knew which pigeon it was by the identity ring on its leg. The astonished hospital staff allowed bird and boy to stay together—a somewhat less remarkable and therefore more credible case of "knowing" (or what we might call inner wisdom of the animal consciousness). There are several authenticated cases of cats and dogs finding their deceased owner's grave and either visiting the spot once, as though to pay their respects, or repeatedly visiting the grave, sometimes even for years.

Animals will follow their human companions to the grave and perhaps even beyond. There is another kind of following called psi-trailing, and from the cases that have been thoroughly researched, there can be absolutely no doubt that this phenomenon exists. One of the most celebrated cases was Lassie, a mongrel dog who was left behind at a small farm in Kentucky when his owners moved out west to California. The dog left the people who had adopted him, and months later, he located his family in Pacoima, California. A cat named Clementine did a no less incredible 1,500-mile journey from Dunkirk, New York, to Denver, Colorado, to find her owners. Another cat named Tom holds the distance record of 2,500 miles or more, traveling from the home place in St. Petersburg, Florida, to San Gabriel, California, to find his owners in their new home.

Several years ago, when a psychiatrist friend of mine was a teenager, he was given a dog by neighbors who moved from Brooklyn to live in Queens, New York. Alan had the dog only a couple of days when it ran off. The owners called a few days later saying that they had found their dog walking up and down their new street in Queens and were so amazed, since the dog had never been there before, that they felt they had to keep him after all.

A similar case of psi-trailing was reported on national news in the spring of 1983. A little mongrel dog, left with neighbors in Colorado, made an incredible journey over the Rockies in the dead of winter and found its owners at their new home in California, where the dog had never been.

A young cow called Blackie was recently sold, along with her calf, at an auction in England (*Agscene,* February 1984). The two animals went off to separate farms. That night Blackie escaped and was found the next morning suckling her calf on the farm where her calf had been taken. Her owner was traced via the auction tag that was still on her back. Out of compassion and amazement, her owner allowed her to stay with her calf. The farms were seven miles apart, and Blackie had never been there before. The bellowing of a calf could surely not travel that distance, nor its scent, unless cows have remarkable senses. Whether Blackie was psychic or telepathic, or had supersensory abilities, is an open question; but no less remarkable is the strength of her maternal instinct and desire to be with her calf.

Without a concrete, physical *explanation* of this psi-trailing ability, many people still remain skeptical, even though these animals and others could be definitely identified by their owners. In fact, the evidence for this animal ability is incontrovertible and irrefutable. Even though we cannot explain the hows and whys (except that one of the major whys is love), what does such evidence mean? What is its deeper significance?

It means first that many animals, possibly your cat among them, possess mental (or psychic) abilities far more sensitive and better developed than ours. It means that animals may be far more aware of or receptive to another dimension of reality that is relatively closed to us: that is, with the exception of human seers, prophets, clairvoyants and our own occasional flashes of déjà vu and precognition.

One of the best publicized cases of ESP in a dog occurred during World War I. An English dog named Prince somehow crossed the English Channel and roamed France until he found his beloved master in the battlefront trenches.

The late Dr. J. B. Rhine from Duke University investigated a number of alleged cases of psychic abilities in pets. The following cases leave no doubt that animals certainly do possess such abilities. One of the most amazing cases of psi-trailing involved a female dog who was adopted by a family at their summer vacation home. By the end of the summer she had presented them with a litter of pups. They were unable to take the dog and litter back to their New York City apartment, so they found a good home for them near their summer home. About a month later the dog turned up at their apartment in the city, an incredible feat since the dog had never been there before. It was some thirty miles from the vacation cottage. The dog was carrying one of her pups. She deposited it at the feet of her amazed owners and then asked to go out. Several days later she returned with another pup and was off again until she eventually had brought her entire family to the apartment. Naturally the people decided to keep her. This story is hard to accept but is one that Dr. Rhine thoroughly investigated and authenticated.

Another of his cases involved a little terrier named Penny. One and a half years after the death of her mother, a daughter in the family came home to make a visit to the graveyard to pay her respects at her mother's grave. She took Penny with her—the dog had never been there before. While the daughter was freshening up some flowers at the cemetery before going to her mother's grave, Penny got out of the car. To the woman's surprise, when she went to her mother's grave, there was the dog lying on it, whining pathetically. No other family member had been there for months, so it is unlikely that there would have been any familiar scents for the dog to detect.

Two other cases studied by Dr. Rhine clearly show that dogs possess ESP. At the moment his master was injured in an airplane crash, a dog became acutely distressed and crawled and hid under the house. There he stayed for several days, during which time his master lay in a coma at the hospital. The dog finally came out from under the house at the exact time that his master regained consciousness.

A dog in Richmond, Virginia, helped save two boys who were out camping. The dog woke the parents by howling in a very distressed way late at night. The parents felt something was wrong and began to worry about their boys. So they drove out to where the boys were camping, about ten miles away. They found the woods on fire, and the fire was moving toward the sleeping boys' tent. They were able to get the boys out in time, thanks to their dog.

For nearly thirty hours in the spring of 1983, villagers on the

Taking an accurate measure of animal intelligence is tricky at best. Can we say that a cat is smarter than a human? We can, and the answer is "yes—at being a cat!"

Photo courtesy Humane Society of Santa Clara Valley/Hutchison

island of Minorca searched for a lost three-year-old child, to no avail. Mayor José Tadeo, leader of the search party, returned to his home two miles away, and his two-year-old Irish Setter greeted him but kept whining and scratching at the door to be let out. The man complied, and the dog got him to follow right to the area where people had been searching for the boy. The dog found the semiconscious boy in a small crevice concealed by undergrowth. How could the dog have known, since it had been at home, two miles away, all day?

There are ancient documents that tell of a time when some people could "talk" to animals, implying that there was some connection between the animal and human consciousness. The later persistence of animal totems, superstitions and such may be remnants of our past kinship in consciousness with animals. How removed we have become, and at how great a loss, as we have become civilized and technologized and increasingly alienated from the inner wisdom of the natural world.

The irrefutable evidence of psi-trailing in cats and dogs humbles me and engenders a deeper respect for the seemingly supernatural abilities of such commonplace creatures. Familiarity with them, if it has not bred contempt, has at least fostered an attitude that animals are at least less intelligent than we. Yet by what yardstick do we measure intelligence? The commonplace question "Are cats smarter than dogs?" reflects the ignorance behind our general attitudes toward animals. The answer is that a cat is smarter than a human being—at being a cat. By the same token we demean animal instincts as being something mechanical, reflexlike and unconscious. Take, for example, the ability of migrant species of birds and even butterflies to travel thousands of miles to specific traditional places that they have never been to before. Even when science explains these migratory abilities as being cued by the positions of the sun, moon, stars and by geomagnetic influences, we must wonder still at how such intelligence is inherited.

The most thoroughly researched and publicized case of homing in a pet was of a Collie named Bobbie. He was lost in Indiana, and his owners had to return to their home in Silverton, Oregon, without him. Somehow he was able to find his way to his owners' new home, some three thousand miles away, in midwinter. This dog's feat was sufficiently well publicized such that people who had given him food and shelter on his long journey home made themselves known. This way, the route taken by the dog was approximately reconstructed.

While we can accept the evidence of animals' abilities to navigate

On many occasions man has benefited from the apparent psychic ability of the animals around him. Cat owners learn to read signals from their pets and have been saved on more than one occasion in crisis situations. *Photo courtesy M. W. Fox*

and either find their way home or migrate to some distant place, and while we can also credit that some animals can psi-trail, we would surely balk at the suggestion that animals may have clairvoyance—that they can see into the future.

During the Second World War, house cats gained a new respect and appreciation in England for their seemingly psychic ability to sense if a bomb or "doodlebug" missile was going to fall too close for safety. Owners would relax while the air raid sirens wailed their alarm and only seek shelter if and when their cats became agitated. Scientists in China and United States are attempting to determine how animals, including pets and zoo animals, are able to sense the advent of an earthquake, sometimes days before its occurrence. Physical explanations, such as the ability to detect prequake shifts in the earth's magnetic field, may be forthcoming. Then such sensitivity would have a physical explanation and would no longer be regarded as "psychic."

I believe (but I still need the evidence to prove it) that feelings, particularly those of love, fear and the sadness of loss, can be sensed, that they, like ideas and mental images, affect a sensitive, receptive mind just as physical vibrations of touch, light and sound affect our sense. Animals are generally more sensitive and finely tuned, so to speak, and have a greater ability to detect slight physical shifts in their sensory field of smell, sight, sound and touch (including ground vibrations). Perhaps they are also sensitive to emotional or psychic changes in their sensory field that to us less sensitive beings are relatively supra- or extrasensory. Just as a bee can see ultraviolet light, and a cat can hear the ultrasonic squeaks of mice, we are blind and deaf to both these extremes of the physical field of our senses. The same may be true of the nonphysical or metaphysical realm, except during those fleeting moments when we are relaxed, meditating or going to sleep, when we are not thinking but are briefly quiet, open and receptive. I like to speculate that millenia ago, when our lives (and minds) were more emotionally attuned, and when we were less preoccupied with planning, rehearsing and reminiscing—in other words, with thinking and talking in our heads to ourselves—we were once free to commune empathetically with other animals, as they may well do with each other today, and "know" intuitively, or psychically, as such knowledge is called today.

I have presented a few anecdotes about some of the wonders and mysteries of animals. What remarkable abilities some have demonstrated in their close affection for people. Irrespective of logical explanation or mystical speculation, the fact remains that animals are not dumb creatures living in a twilight world of partial awareness. Nor are they unfeeling, instinctive machines. Suffice it to say that if there is a life hereafter and we continue to demean animals and treat them without compassion and respect, then there will be an accounting. And if there is no life hereafter, let us open our hearts and minds to them, for they will indeed enrich our lives, sometimes with more than love and companionship. They may open the door for us to a different reality, the natural, phenomenal world, of which we, because of our state of mind and life-styles, are no longer a part. We are more ignorant than they in this respect, as so often we are more ignorant of our pets than they are of us!

An essential ingredient in many of the remarkable "psychic" feats of animals is love: their strong emotional attachment to their owners. The importance of proper rearing and socialization early in life, which help establish a strong emotional bond, cannot be overstressed and is the subject of the chapter that follows.

The first lessons of life a kitten learns are from its mother. When a well-adjusted kitten is acquired at the optimum bonding age, there is every chance that it will grow into a Supercat—and will be a satisfying companion for its entire life.   *Photo courtesy Joseph R. Spies*

7

Raising a Supercat

THERE ARE THREE CRITICAL FACTORS to consider in raising a "supercat." These are termed *socialization, early handling* and *environmental enrichment*. There is a so-called critical socialization period that is the best time to acquire a kitten.

This time is between eight to ten weeks of age, since research studies have shown that this is when a kitten most readily "bonds" to people after it has been weaned. Regular handling well before weaning is also beneficial. What socialization does is make the animal emotionally attached to you, and this will make it more trainable. If socialization is delayed, say until the kitten is three or four months or even older, it will not be sufficiently bonded to you to make for a close relationship. Also, this *primary* bond between the owner and the cat must be well established during this early critical period for the *secondary* bond (liking and trusting other people as well) to develop.

Socialization consists of many important forms of interaction between the young animal and its owner. The process is essentially the same for all young animals, including cage birds (such as parakeets and parrots), cage mammals (rats, gerbils and hamsters), as well as cats, dogs, horses, sheep and others. To some extent, socialization also takes place in such creatures as snakes and other reptiles like turtles and lizards; they become habituated to human handling and lose their natural fear of people.

Socialization first involves *exposure* to humans, so that the young animal becomes accustomed to human presence and to the behavior, sight, sounds and smells of people. Second, through regular handling, it becomes accustomed to being held, restrained and carried. Those species that are *contact species* and naturally enjoy being licked, groomed or preened by their own kind will learn to enjoy being stroked or petted by a human being. But for an unsocialized animal, human contact evokes fear and avoidance.

Early exposure to human touch (and to quiet "grooming-talking") establishes a *pleasure bond* so that later in life the kitten will enjoy and regard as a reward a stroke or kind word of praise from its owner. Petting produces a marked decrease in heart rate, which is a sign of general parasympathetic nervous system stimulation. It is through such stimulation that the pleasure bond and social attachment with an alien species (in this context, with humans) can be established early in life.

Socialization includes other modes of interaction. Caretaking, providing food, water, shelter and security, are essential *parenting* functions. They mimic the activities of the infant animal's natural parent so that the animal becomes attached to and dependent upon its human foster parent.

As a natural parent may sometimes play with, explore and discipline its offspring, so the human "parent" should engage in such activities. Regular games and romps, tugs-of-war, hide-and-seek, and play-fighting are bonding activities: Those who play together stay together. Exploring the environment is an essential part of growing up. With the security and instruction of the natural parent, the young animal learns about its environment and through observational learning acquires knowledge from the parent essential for its survival, such as where to hunt or find a safe hiding place and related skills. The human foster parent should do no less with a kitten. An animal constantly confined to a cage, kennel or indoors will be relatively experientially and environmentally deprived. Its early curiosity (*neophilia*) of novel objects, which naturally leads it to explore its world and so acquire knowledge, will be inhibited by fear of the unfamiliar (*neophobia*). Early environmental enrichment—providing paced increments of experience by exposing the young animal to all kinds of novel stimuli *always under a protective parental eye*—is part of the socialization process. However, a kitten should never be allowed to roam free and unsupervised or be encouraged to wander alone outdoors. Under the security of the parent, the infant individuates and gains its independence. Overprotective parenting—not allowing the animal to investigate unfamiliar situations and objects, whether innocuous or potentially dangerous—can be as bad as raising the kitten (or child) in a restricted environment.

Discipline is the other essential aspect of optimal socialization. Most mammals will discipline their offspring when they step out of line, biting too hard during play, or not respecting the social status of their parents and the "rights" of others.* It is through effective parental discipline that a socially well-adjusted individual develops. Overindulgence and permissive rearing is all too common in pet owners and usually results in a cat that is disobedient, always gets its own way and does not take well to corrective training as an adult.

In good animal (and human) parenting, there is a subtle balance between discipline and affection: the latter is not conditional upon conformity, although good behavior may be rewarded with affection. There is also a subtle balance between parental protectiveness, infantile dependence and maturing independence. Overprotectiveness can interfere with individuation/independence, and the denial of autonomy will limit the development of intelligence.

The basic temperament of the animal can be improved through optimal socialization and careful handling. For example, a kitten with a timid disposition, if correctly handled, could be helped to become a more stable and adaptable animal by "buffering the genotype."

The earlier such a kitten can be identified, the sooner it can be helped. A person used to handling kittens only a few days old can "feel" differences between littermates. Potentially timid, unstable kittens are tense while being handled and may vocalize distress when gently turned over. Once disturbed, they will take longer to settle down than more stable littermates. A stethoscope could be of great help; unstable puppies as young as three or four days of age generally have a *slower* resting heart rate than their littermates, the most outgoing of which as a rule have higher resting rates. Kittens may show a similar correlation between heart rate and temperament, but this has not yet been fully researched.

If the kitten is taken from its mother too early, say between the fourth and fifth week, it can become too people-oriented, and this can also lead to problems, such as fear of or aggression toward its own

* "Rights" include the right to sleep in a certain place, to be left alone (that is, to privacy) and to eat in peace—rights that are often not respected by owners or by their children.

kind, and even refusal to breed. Some early-weaned kittens become tail or paw suckers later in life or suck and chew on blankets. I have seen many cats who have had plenty of socialization to people but have not had what I call *environmental enrichment*. There is a critical period for socialization to people, and there is also a later critical period for learning to cope with unfamiliar environments and novel stimuli. Any kitten kept cooped up in a cage, or house, especially during its early formative months will often show fear when taken into unfamiliar environments or situations, such as being in a car or meeting strangers.

With an especially outgoing kitten, one with an innately stable temperament, recovery from this early deprivation is usually rapid if the kitten is not kept confined more than three or four months. But with an innately timid or fearful one, it may be very difficult, or even impossible, to cure this condition. This problem can be prevented. I advise people—as soon as their young kitten has received protective vaccinations—to take it out with them everywhere they go, into all sorts of situations, in the car, into the field (never alone), and so forth, and to introduce the kitten to as many different people as possible. The young animal's experience of the world must be allowed to expand naturally during the first few months of life; otherwise its potential may be thwarted. A kitten that is afraid of novel things—a reaction called *neophobia*—will be difficult to handle in strange situations and will be a poor learner. Its IQ will suffer because it will be too afraid to explore and investigate new things and learn in the process.

There are benefits to be derived from *early handling,* which can begin at birth. This entails simply picking the kitten up, repeatedly turning it around and upside down, and stroking it for a few minutes every day. The animal can also be placed for one or two minutes on a cold surface (a linoleum floor is ideal) to arouse it and to induce very mild stress. With this early handling from birth to four weeks of age, accompanied by a little stress, you can produce a kitten that is physiologically more resistant to stress in later life.

We have discussed what *should* be done during the early critical weeks of development. Now to detail what should *not* be done during this period.

In order to avoid extreme fearfulness and submissiveness, it is important at this time not to be too controlling with the new kitten. Training should be done with a very light touch. In particular, *no* disciplinary training should be done between the eighth to tenth weeks, since this is a very *sensitive period* in a kitten's life. This is an espe-

The actions of a well-adjusted kitten often reveal that it was properly socialized at an early age with plentiful handling and human contact during its critical stages.

Photo courtesy M. W. Fox

cially sensitive time because it is when kittens are most likely to develop an avoidance response if subjected to physical or psychological trauma. The aim at this time should be to give the kitten a sense of confidence and *control*. Kittens develop a sense of control and mastery over their environment when they are given a chance to play, explore and learn. When they are kept in a cage all the time, where they are provided with food and water but no stimulation, they are in a totally controlled and controlling environment, in which they don't develop a sense of confidence. And if the kitten is afraid of new experiences, it is not going to learn. Fear gets in the way of learning and greatly limits the animal's potential.

There are some other special socialization problems that we need to consider. Cats raised by women can become afraid of men; cats not raised with their own kind or denied opportunity to interact socially with other cats may become fearful or aggressive toward them later in life. Cats who do not have contact with children during the socialization period may be difficult to handle around children later and show intense fear or, more rarely, aggression.

The basic lesson is that during the socialization period, broadly defined as extending from the first through the third month of life, kittens should be exposed to all sorts of positive experiences and all sorts of people, with care being taken to avoid threatening or stressful experiences, especially during the second month.

SHYNESS

There are different kinds of shyness that must be recognized before an appropriate way of dealing with the problem can be tried out. First, we must distinguish between shyness and aloofness. Many cats that are extremely friendly, attentive and responsive as kittens become more standoffish and independent as they mature. This is a quite natural phenomenon and shows that animals do have personalities that take time to mature, just as ours do.

Some cats, however, as they mature become increasingly fearful toward strangers (human or animal) and hide when strangers approach. Yet they are never this way with their owners. This kind of shyness is very common in cats who spend most of their time indoors and rarely meet other animals or people.

Such animals should not be forced to make friends. This will only frighten them and make things worse. My advice is to let them be. However, if a new person (spouse or friend) or another cat or dog is coming to live in the house permanently, steps must be taken to help the shy cat overcome its fear. Putting a little of the *same* perfume or after-shave on people's hands and ankles and animals' heads and backs sometimes helps, since it is the stranger's odor that often triggers alarm in shy cats. A few dabs every day for seven to ten days is worth a try if the cat hasn't started to show signs of accepting the human or animal newcomer during their first ten to fourteen days together.

If the odor treatment does not work, then a course of treatment with Valium is often effective. After three or four days of treatment, the cat should be kept in the same room as the newcomer and not allowed to run away and hide. If there is no significant improvement after ten to fourteen days, then treatment should be stopped. Cats that are resistant to these treatments are sometimes shy by nature, which is an inherited problem.

Valium treatment (under strict veterinary supervision) is also effective in some cats who suffer from shyness of a new place rather than shyness of strangers. This neophobia, as it is called, often occurs when a cat is adopted into a new home or when owners take their cat to live in a new house. A common sign of emotional distress in these cats, as well as those who are shy of strangers, is to become unhousebroken.

Shyness in cats can be inherited. So one of the best preventives is not to knowingly breed and produce offspring from shy parents.

Some kittens are by nature on the shy side. One way of helping them become more extroverted and to gain self-confidence is to play with them frequently. Giving a kitten a strip of fake fur or a feather on a string to chase and catch as you pull the string to animate the "prey" will help the kitten enjoy your company and become more human-oriented as it matures. And a regular grooming—even massage, as I describe in my book *The Healing Touch*—for all kittens will help prevent shyness later in life because such contact establishes a bond of affection and trust.

The foregoing early handling, socialization and environmental enrichment procedures are the three golden rules of creating a *super-cat*. This is not meant to imply that it is possible to improve upon nature, but rather that by bringing out the best qualities and buffering or compensating for negative attributes, one may create an adaptable and intelligent animal. Studies with infant rats, kittens and puppies given early handling, socialization and enrichment have shown some of the following attributes: greater resistance to physical and emotional stress; increased resistance to several diseases, including leukemia and induced cancer; enhanced learning abilities because of greater emotional stability; a generally outgoing character, which means the animal is more likely to learn by being highly inquisitive and exploratory.

SOME GENERAL EFFECTS OF DOMESTICATION

The brains of our pets are smaller than those of their wild counterparts. Generations of domestication and being raised in a relatively unstimulating environment in cage, house or suburban yard have resulted in a reduction in brain size and intelligence in our pets.

Over a hundred years ago, Charles Darwin observed that the domesticated rabbit has a much smaller brain than a wild rabbit, a fact he interpreted as due to a deprivation of natural outlets for the development of normal instincts and behavior patterns.

More recently, research scientists have demonstrated that laboratory rats have smaller brains and are poorer learners than littermates that are raised together in enclosures "enriched" with novel toys and manipulation.

It is known that wolves have an average brain size one sixth larger than that of a dog of the same physical proportions, again confirming that domestication can have a marked influence on brain development. (Domestication can also change other anatomical structures, the intestinal length of house cats being found to be longer than in wild cat species of comparable size, possibly because they are fed more cereals and other high-fiber ingredients.)

Rats, mice, dogs and captive-raised monkeys, if provided with a rich and varied environment as they are maturing, will be better problem solvers—that is, more intelligent—later in life. Wild and domesticated strains of rats born and raised under the same conditions show marked differences in behavior and intelligence. Wild rats, even when bred in captivity, show more complex behavior and are better learners than the bland white rats who have been domesticated over hundreds of generations of selective breeding for docility.

What these studies disclose is that two interrelated processes are involved in the gradual decline in brain size and intelligence in our domesticated companion animals.

In most places, a free-roaming cat is exposed to many serious hazards. Owners wishing to prevent such exposure can exercise several options. This enclosed patio allows the cats complete access to fresh air and sunshine while keeping them safe from the perils that can otherwise befall them. *Photo courtesy M. W. Fox*

If space and conditions permit, an attractive outdoor enclosure can also be used to furnish an outdoor environment for cats without putting them at risk from accidents, other animals or hostile people. *Photo courtesy M. W. Fox*

First, in a relatively protected domestic environment, there is a relaxation of natural selection pressures that favor intelligence and alertness. A domestic pet need not be intelligent to survive, while in the wild the unintelligent, the poor learners and the soporific would not live long. Nor would their attributes be likely to be passed on to any offspring, since they would die before they reached sexual maturity.

In the more protective domestic environment, natural selection for survival-enhancing behavior is therefore reduced, and there has been a greater emphasis placed upon tractability, docility and submissiveness toward humans. The general levels of alertness, activity and curiosity (especially arousal by novel stimuli) have also been reduced—and in some cats to the point of making them "sweet nothings" that are relatively insensitive to or unaware of much that is going on around them.

While there has not perhaps been a deliberate attempt to lower intelligence in domesticating animals, in selecting for a more docile and easygoing animal, arousability, alertness and curiosity may be so reduced that the animal learns very little except when instructed or trained.

Neophilia and neophobia—curiosity and fear of novel stimuli and unfamiliar events—are qualities of the wild animal. They are related to the animal's general state of arousal and attentiveness. No one wants a relatively hyperactive and hyperalert cat that must explore everything new in its environment or flee from anything unfamiliar and intimidating.

In breeding out or reducing these "wild" traits, the domestic cat is made easier to handle and control, but since it is not going to explore its environment with the intensity and thoroughness of its wild counterparts, it will acquire less knowledge. Its IQ will be low as a result of selective breeding for desirable domestic qualities.

A second reason why many pets have low IQs is that they are raised in a relatively bland, understimulating and monotonously predictable environment. Life in a cage, a small apartment or suburban back yard can be so experientially depriving that the animal's potential is never fully developed.

In addition to varying degrees of experiential and environmental deprivation, emotional influences can also get in the way and further restrict the animal in learning new things and, therefore, developing its IQ. These same emotional and motivational influences can also interfere with IQ tests and "educational enrichment." An animal that is afraid or anxious will not perform well. Similarly, one that is basically timid, or too headstrong, or not sufficiently motivated by hunger or curiosity (depending upon the nature of the test) may also give an inferior performance.

Aside from carefully breeding cats to be more alert and outgoing, yet emotionally stable, to improve their IQs, we can also provide them with a more natural environment that will help bring out their potential and make their existence more enriched and fulfilling.

In conclusion, raising a kitten is not unlike raising a child. But love alone is not enough. Knowing the developmental stage that your kitten goes through and the critical times during its development when special care and attention are needed can help bring out the best in your feline companion.

POSTSCRIPT

Can "Wildcats" Be Tamed?

A common question from many prospective cat owners concerns the suitability of adopting a "wildcat," such as a kitten from a litter that has been born in the woods or raised by a mother cat who lives in the wild or on a farm. Indeed, a surprising number of kittens and young cats come from the wild, and many finish up in animal shelters for adoption. Provided the kittens are handled regularly early in life, especially before they are eight to ten weeks of age, they will indeed make wonderful companions. Research studies have shown that truly wild kittens can make wonderful house pets if they are adopted young enough. The longer they live away from humans, the more fearful they become. Heredity can also play a role. Kittens from very wild or fearful mothers may, as they mature, become less friendly and more aloof, which in part explains the apparent temperament changes in many cats as they mature. Recent research has also revealed that the temperament of the father can also influence the behavior of kittens, even though the father never has any contact with them after they are born. Friendliness toward humans has now been partially explained in terms of who the kitten's father was!

A companion cat, unlike other pleasure animals, brings more of the natural world into our lives. Even highly cultivated purebreds like the Siamese shown above is considerably closer to wild relatives like *Felis libica* (below) than domestic dogs are to any wild canids.

Upper photo courtesy HSUS/Hudson; lower photo courtesy M. W. Fox

8

The Benefits and Responsibilities of Cat Keeping

AT ONE TIME we lived not simply close to nature, but *with* nature, like all the other animals. But today most of us have no real connection in our everyday lives with nature or with wild animals. Our parks and gardens are merely facsimiles of all that was once wild. And the few "wild" creatures, such as birds, rodents, snakes, lizards and insects that live close to us (and which we often demean as "pests"), are the last survivors of the animal kingdom that figure into our daily lives.

Like the tamed landscapes that we have created, so are our domesticated farm and companion animals tamed, docile and transformed facsimiles of their wild ancestors.

But there is one exception: the cat. Unlike most breeds of dogs, the average cat is much closer to its wild relative, the desert cat, *Felis libica,* than the dog is to its purported wolf ancestor, *Canis lupus pallipes.*

Because of its naturalness, self-reliance, independence of will and wild spirit, the cat appeals to those persons whom I call "cat people.'' The natural cat loves you on its own terms and is quite capable of taking care of itself and of living independently as a hunter in the wild.

The average cat is wilder and more natural physically and psychologically than the average dog because it has been domesticated for a much shorter period of time, approximately six thousand years—first by the early Egyptians—compared to an estimated fourteen thousand years for the dog. And during most of this time, the cat was simply tamed, socialized to humans, but not changed significantly in other ways because it was kept to be a hunter, especially to keep rodents away from households, grain and food-storage areas.

We do not see such extreme variations as the Chihuahua or Great Dane in cats for these reasons. But as breeders begin to selectively breed cats with mutant traits, such as brachycephaly (short faces), dolichocephaly (long faces), large and small stature, folded ears, long and thin bones, unusual eye and coat colors, curly hair, hairlessness, taillessness and a wide variety of other traits, we will in a few decades have as many recognized varieties as there are various breeds of dog today.

All ''cat people,'' as distinct from ''cat fanciers,'' share my concern that many of these new varieties will suffer from inherited health problems that are so common in purebred dogs as a consequence of selective breeding for certain traits and the use of inbreeding. This can be avoided and the welfare and rights of future generations of cats better insured by cat fanciers' not producing structurally, functionally and otherwise abnormal cats—and by prospective cat owners' choosing healthy, natural cats.

Some cat lovers go a little too far, however, believing that if they are going to keep a cat, it should be not only a natural cat but also able to live as natural a life as possible. And that means being able to roam free outdoors, hunt and kill wild animals and also breed with other cats and have offspring. While I appreciate the sentiment of such cat owners, I regard them as irresponsible. Free-roaming cats can get into fights with other cats, which often result in severe bite abscesses; they are also likely to pick up infectious diseases such as feline AIDS, distemper (panleukopenia) and leukemia from other cats, rabies, toxoplasmosis and plague from wild animals, and bring home fleas, lice, ticks and mites. Many free-roaming cats are killed by automobiles and not infrequently by free-roaming dogs and hunters. In some areas they also fall prey to bait-poisoned rodents they eat, to traps set to catch wild animals for their fur and are stolen and sold to research laboratories or used to train pit bull types to fight and kill.

Free-roaming cats also get lost and starve to death or die from infections. Many will kill wildlife, even bringing dead or injured prey home. Wild animals have enough problems surviving as it is, and cat people who respect the rights of all creatures keep their cats indoors to prevent such needless killing. An exception may be made for "working" cats that control rodents around farms and warehouses. But these cats should be provided with veterinary care when needed, given vaccinations and neutered to control their numbers, three basic rights that are rarely respected.

Free-roaming killer cats also compete with and displace native wild predators like the fox, marten, weasel and wildcat, sometimes even transmitting infectious diseases to them.

I respect and love cats, but one thing that does enrage me is seeing them carrying dead birds, squirrels and other wildlife that they have killed. Only too often, since they are home-fed house cats, they don't eat what they kill, so the death of these creatures is pointless. One could argue that it is right to let cats follow their natural instincts and that it is an inhumane violation of their rights to confine them indoors when they want to go out to hunt and kill. But what of the rights of wildlife that have a hard enough time surviving, since so much of their habitat is being taken over and destroyed by people? That cats help keep certain urban and suburban wildlife species "in balance" (like rats, mice and pigeons) is a flawed justification for letting them out to kill, since they are generally nonselective. They kill whatever they can, from songbirds and helpless ground-nesting birds, such as pheasants and quail, to harmless snakes, lizards and young rabbits and squirrels.

One evening I saw a neighbor's cat carrying a dead squirrel in its jaws. I asked the cat's owner, who was in her garden next door, how she felt about allowing her cat to roam free and kill wildlife. She immediately took offense, thinking I was telling her how she should keep her cat, and demanded that I mind my own business. But it was my business—I was concerned for the squirrel and other wild creatures that her cat and a half dozen others in the vicinity were killing constantly. She blamed one of the other cats for encouraging hers to hunt, and I pointed out that cats hunt alone and don't need others' encouragement. Then she said she hadn't seen her cat for two days and was worried. I suggested that it might be going wild, feral, and that the more it got a taste for the outdoors, the harder it would be to keep it indoors. When kept indoors, her cat expressed its frustration by spraying all over the place, even though it was neutered.

My advice to people with killer-cats that can't tolerate being indoors all the time is to not let them out except long after dawn and bring them in well before dusk, since cats hunt most efficiently around dawn, dusk and through the night. They should have a bell attached to an elastic collar to warn their prey of their presence.

A better alternative is to try training the cat to enjoy being outdoors in a pen or on a line and harness (provided the yard is dog-proof). Some enjoy walks on a leash and harness, but be warned—if frightened, the cat may climb up you for protection and injure you with its claws.

Prevention is the best medicine, and by this I mean starting out right from kittenhood and never letting your cat outdoors. Life indoors can be satisfying for cats provided with a companion cat, carpeted window shelves to sunbathe on and from which to look outdoors. Cats don't usually get frustrated when kept indoors until they have had a taste of the outdoors. And remember, it is safer indoors. If you care for your cat, you should care for all creatures, and it is your responsibility to protect wildlife from your cat and to protect your cat from outdoor hazards. The woman whom I offended when I confronted her about her killer-cat loved her cat. But love is not enough: responsibility and understanding make a cat lover into an animal person, one who respects not only the rights and needs of his or her own animal companion but also the rights and interests of other creatures wild and tame.

But how safe is your home for your cat? There are indoor hazards to look out for if you have a cat or other household pet.

One of the first cases that I had to treat when I entered veterinary practice was a young cat with a high fever and a telltale piece of cotton thread coming out of its rear end. I suspected the cat had swallowed a needle with the thread attached and the needle was now penetrating its intestines. X rays confirmed my diagnosis, and we were able to save the cat after extensive surgery to remove the needle and repair its intestines.

This kind of problem is household hazard number one for cats. Small objects left lying around that the cat may chew and swallow can mean death. Never leave anything within your pet's reach, especially if it is a playful and very ''oral'' kitten. Objects small enough to be

Many cats enjoy walking outdoors on a leash. If your cat is one of them, you are fortunate to enjoy its company in this way. Have your cat fitted with a proper harness for outdoor excursions. A harness is safer and more comfortable for a cat than a collar. *Photo courtesy M. W. Fox*

Allowing cats total freedom is a common practice that will probably never die out. The best authorities counsel that cats are indoor pets and should not be allowed to roam. Anyone whose outdoor cat has ever disappeared, been killed, injured or presented its owner with an unexpected litter should understand the wisdom of the indoor life for a beloved housecat. *Photo courtesy M. W. Fox*

swallowed or likely to break when chewed to the point where small bits could be eaten represent considerable danger.

Many times I have treated kittens who have found a piece of string or thread to play with. They start to nibble on one end and nibbling leads to chewing and chewing to swallowing. Once the material is in a cat's mouth, it is especially difficult for the animal to remove it because the rasps on the cat's tongue point backwards. So the cat keeps on swallowing. A twisted mass of string or other material can cause an intestinal blockage and death. Only too often there's a needle on the end of the thread, and this will perforate the intestines and lead to peritonitis. Few pets can be saved surgically when this happens. So please, during and after sewing, keep an eye on all threaded needles and be sure to tidy up completely.

One should be especially alert with new kittens because they are so active, inquisitive and likely to find things under or behind furniture that you might have overlooked. Clean up thoroughly and give your pet safe toys to play with: rubber, cloth or rawhide toys that won't break or splinter and that are too large to be swallowed. If not provided with any toys, your pet may well try to find its own and get into trouble.

Sometimes a cat will seem to be choking and will paw at the side of its face in desperation, but you find nothing stuck in its throat. Look carefully at the roof of the mouth: bone splinters, a piece of pencil or other object could be stuck in the teeth across the roof of the mouth.

Frantic pawing of the mouth could also mean that your pet has injured its mouth with an irritant household chemical, or it may have burned its mouth after chewing through an electrical cord.

Electrical burns are not uncommon in kittens. Very often they go into shock. To prevent this, be sure that all electrical cords are out of reach, either hidden behind furniture, concealed under a carpet or run along the ceiling. Providing kittens with safe toys and plenty of attention and play to reduce their drive to explore and get into trouble when they are "bored" are major preventive measures.

One often overlooked hazard for cats is getting literally hung up by one or more claws on curtains and rugs. Some materials, because of the weave, or loops of thread, can snare a cat when it claws into it. Claws may be torn and considerable injury result as the cat desperately tries to extricate itself. Regularly trimming the claws or providing one or more carpet-covered scratching posts are good preventives.

A wide variety of household chemicals—from disinfectant sprays and floor cleaners to deodorants, furniture polishes and insecticides—can be hazardous to pets. Many are hazardous to human health as well. Poisoning may be acute—as when a kitten gets into a closet and eats a packet of garden insecticide or rat bait—or chronic. Cases of chronic cumulative poisoning may occur with constant exposure to certain household disinfectants and floor and furniture polishes (which some pets will lick up or pick up on their feet and subsequently lick off). Sometimes contact-allergies develop. The safety measures here include keeping *all* poisonous substances out of the house, locked up in the garage and out of reach of both pets and children. Better still, don't buy or use them. Far too many dangerous garden insecticides and weed killers and indoor pesticides are in common use. Many of these can cause acute sickness if accidentally consumed or inhaled. Worse is the chronic cumulative poisoning that can cause cancer and birth defects.

All household chemicals—floor waxes, cleaners, disinfectants and so forth—as well as human drugs, including aspirin (which can kill cats), should be kept securely closeted and totally inaccessible to your pet.

Look around your house and think what other hazards there might be—what is left to consider? Yes, there are more potential hazards. The first is plants. Certain indoor plants can be poisonous when eaten, or extremely irritant to your pet if accidentally chewed. Look out for daffodils, castor bean, certain ferns, autumn crocus, lily of the valley, cannabis and dumb cane or dieffenbachia.

Keep all plants out of your pet's reach, either on high ledges or window shelves, or in ceiling-hangers. Try growing a little fresh grass for your indoor cat to satisfy the craving for green roughage. Many cats relish it and if they have a supply, will leave your plants alone. Another hazard is related to human behavior around the house. Small pets easily get trodden on or trapped and squashed under or in the door. Always, with a new kitten, open and close doors carefully and slowly. Walk cautiously to avoid stepping on your pet. High heeled shoes are lethal. One of my first emergencies was a Toy Poodle that had been impaled under its owner's stiletto heel. All cat owners should use window screens. An open window can mean a lost cat or a dead or injured one on the ground if you live in a high-rise apartment. A cat intent on birds outside might be oblivious of height and make a fatal leap.

Garbage is a household pet hazard. Cats will raid the garbage can, especially when something inside smells good. Be careful not to place any food that might be attractive or harmful to your pet in an unsecured garbage can. Rotting food can cause severe gastrointestinal distress, and chicken or other bones could kill your pet. So keep the trash can securely closed or out of your pet's reach in the closet under the sink.

Finally, there are some seasonal hazards to consider. Christmastime can be hazardous to a pet's health in many ways. Gift wrapping, decorations, tinsel and Christmas tree light extensions can be chewed or swallowed and could mean an unhappy Christmas for both pet and family.

I consider both yard and car an extension of the home, and both can be hazardous to a pet. Look out for slug bait and any chemicals—weed killers and insecticides—you may have recently sprayed. Antifreeze in the garage is a potent poison. It has a sweet flavor and is attractive to cats.

Sometimes one's cat (or a neighbor's) will crawl under a car and lie near the engine to keep warm. I have seen more than one cat mangled by the fan and belt when the car was switched on and no one knew that a cat was under the hood.

Aside from the natural adaptability of cats, and bearing in mind the virtue of owning two cats rather than one, there are many benefits to be derived from keeping them as companions.

A friendly cat is a catalyst for conversation and a facilitator of human social interaction when visitors come by—or a new girlfriend or boyfriend, for the single person. Most women respect and prefer men who are sensitive and gentle, and it is certain that a man who shares his apartment with one or more cats and loves them will be more able to demonstrate these qualities than one who has no cats. Not all people like cats, however: some are afraid and others simply don't like animals. A cat can help one avoid the latter and win over the former.

Recent studies of pet owners have shown that having a companion animal such as a cat can greatly alleviate a single person's loneliness and bouts of depression. Cats, because they are less demanding and require less attention than dogs, are perfect companions for the elderly and the infirm. Having a cat or two for company when one is more or less confined indoors, or comes home from work to an otherwise empty house or apartment, certainly lifts one's spirits. Having another living thing to talk to, stroke and love that is responsive and affectionate helps one feel more secure and fulfilled. Studies have shown that contact with a pet also lowers a person's blood pressure, which is part of the beneficial relaxation-reflex response. This has been proven to reduce significantly the incidence of heart attacks.

It is little wonder, then, that cats are good for our physical health, since they can contribute so much to our emotional well-being. Yet, regardless of these benefits, I am opposed to people keeping a cat, or any other animal for that matter, as a pet if they are not responsible owners and do not respect the cat's basic rights. These basic rights are the right to a wholesome and healthful diet; proper veterinary care when needed (no home doctoring), and routine health care and maintenance examinations and vaccinations; proper care, empathetic understanding, humane treatment and respect as a sensitive and intelligent living being with interests, needs and emotions.

One or more kittens or cats in the home are of especial benefit to children, who can learn these basic rights from their parents, which will help foster in them a reverence for all life. This will make them more compassionate and understanding as they mature, sensitizing them to the responsibilities of future parenthood. And, it is hoped, growing up around cats will expand their circle of compassionate concern for all creatures whose rights are barely recognized and too often violated by our "civilized" society today.

In other words, the ultimate benefit of keeping cats is not for people's sake, or for cats' sake, but for the sake of all creatures great and small. Children raised with one or more animal companions are far more likely to be concerned as adults about the welfare and rights of all creatures, compared to those who are deprived of the kinship of animals. Of course, the influence of parents and teachers is important in this regard.

Cats are not commodities, status objects, toys for children, tools for research or promotional gimmicks for a consumer society. They are other beings who can grace and enrich our lives and who are worthy of equal and fair consideration, if not reverence and unconditional love.

Even though cats as we know them are not wild animals, they should not be forced to live in a bland, unstimulating atmosphere. What we do to improve life for our pets will be reflected in the satisfying pets they can be. *Photo courtesy HSUS/Dantzler*

9

How to Make Life More Natural for Your Cat

As I DRIVE INTO THE CITY to work each morning, I often wonder just what is natural anymore. Small parks with their pigeons, starlings and sparrows, smoggy trees and shrubs seem like remote little islands of nature amid a sea of concrete. Even the parks aren't natural. Many of the birds are alien species, introduced years ago from Europe, and some of the shrubs and trees aren't indigenous to the region, either. Yet, without its parks, Washington, D.C., and other large cities would be even less conducive to our general well-being. The animal in us seems to need a semblance of nature, a facsimile at least, for our psychological well-being. The same is true for our pets. They, like us, have certain instincts and basic needs that can be frustrated if we do not provide some natural outlets for them. Mental and physical well-being are closely linked with how conducive the environment is to satisfying such needs.

Research has shown that if an animal is raised in a deprived environment, such as in a small cage in a zoo or laboratory, its basic drive to explore and therefore to learn is either inhibited or frustrated.

The development of its brain may then be adversely affected by being raised in a bland, unstimulating environment. Being deprived of various outlets and things to do and to explore, it may become a "zombie," obese, infertile, glassy-eyed and unresponsive, or because of the deprivation-frustration, it may become hyperactive, pace to and fro in its cage and overreact to any changes in its cage environment or daily routine. If it is caged with other animals, more obvious anomalies may develop, especially when the additional element of crowding stress is evident. Schizophrenialike withdrawal, excessive aggressiveness, infertility and increased susceptibility to certain diseases have been documented in animals kept together with no opportunity to get away from each other for a while. In contrast, animals that are caged alone for extended periods may show signs of social and emotional deprivation and lick or groom themselves excessively, eat or drink excessively, develop stereotyped pacing or running patterns, and even become infertile (because of lack of adequate social and sexual stimulation). Sometimes they will use some object in the cage, such as a piece of food or a water bowl, as a substitute for a sexual partner or for prey to chase and catch, or else they will use one of their own appendages, such as a tail, as something to chase or even to nurse on for comfort, and this may eventually lead to self-mutilation. When a companion animal is introduced to one that has been caged alone for a long time, it may either ignore it or attack it viciously and even kill it. Conversely, because of overattachment, if one caged animal loses its companion, it may go off its food, groom itself excessively and mutilate itself or even die from depression. A mother monkey or ape may become so stimulated by its infant that it grooms it too much and handles it so much that it may die because it never has time to nurse; if it does survive alone in the cage with its mother, they may become overattached, and the infant's development toward independence may be retarded.

These extreme examples of what can happen to cage-raised animals in zoos and laboratories may hardly seem relevant to caring for our pets, yet I have seen similar and sometimes identical problems in house cats suffering from the effect of living in a relatively unnatural environment, like a house or apartment.

I will describe in detail some of the problems that can arise from keeping a cat in an unstimulating and unnatural environment and also offer several remedies by which the cat's world can be made more natural and therefore more conducive to satisfying its basic needs.

Often the housecat's access to the outdoors is through an open window. It is essential to protect our pets as much as possible, and in the case of cats, keep them indoors. While most cats enjoy being outside, it is impossible to know when a given excursion may be your cat's last.

Photo courtesy M. W. Fox

Self-grooming is a normal part of every cat's hygiene. However, excessive grooming activity can indicate social and emotional withdrawal as consequences of extreme isolation.

Photo courtesy M. W. Fox

Many of the symptoms seen in captive wild animals are less obvious, and fortunately sometimes even absent, in our pets because of domestication. Through domestication (selective breeding over generations) cats have become generally more subdued, docile and tractable than their wild counterparts. This means, up to a point at least, that they can cope well with a reduced level of stimulation in that they will not become hyperactive and are less easily bored or frustrated. Conversely, when they are stimulated by something novel and unfamiliar, unlike a wild animal, they will not usually overreact. Some pets will not even notice a change, say, in the arrangement of living room furniture or a new tree in the garden, but a wild captive wolf or bobcat would notice at once.

While our domestic animals have been tuned down somewhat in contrast to their wild counterparts in order to make them more adaptable as pets, they may still suffer from certain deprivations, and therefore, whenever possible, we should endeavor to make life as natural as possible for them.

OBESITY

It has been estimated that 30 to 50 percent of our pets are overweight. This can lead to other complications such as circulatory anomalies, heart failure, diabetes, infertility (if they are not already neutered), arthritis and possibly accelerated aging. Steps toward rectifying these problems are part of making your pet's life more natural. This includes a more natural diet and feeding schedule: don't overfeed or feed too often, and do not give your cat an excess of table scraps or a diet high in calories. Feed a scientifically formulated complete and balanced diet. Also provide regular exercise through play (chasing and retrieving a toy, hunting and catching a ''mouse'' toy on a string).

BASIC DRIVES AND INSTINCTS

Much can be done to make life more natural for your cat once its basic needs and instincts are understood. By the same token, life will be much easier for you, since coping with a frustrated pet can be difficult and having a naturally behaving pet can be an enjoyable and educational experience.

First, the basic feline sex drive, especially in males, needs consideration. Generally cats adapt better when they are neutered, and in a way, such treatment is in fact making life more natural for them. I say this because in domesticating cats we have increased their sex drive. Male wildcats only produce sperm during a short one-to-two-month period each year, while our pet cats are constantly potent. It is more "natural," therefore, to have them neutered, and they will be less sexually frustrated and will not want to roam the neighborhood to find mates and to fight with rival males.

The hunting instinct is another deeply ingrained behavior that is frequently frustrated. Pet cats may leap at their owner's legs after waiting in ambush behind a sofa for their "prey." The remedy here is frequent play sessions with suitable prey-substitute toys for the cat to catch and "kill." It takes only a little time and imagination to come up with appropriate toys to keep a cat happy and active in this regard. Many cats also enjoy carrying their "prey" around with them, collecting and hiding them in various parts of the house.

Cats have an instinctual need to climb up on things. They enjoy climbing trees, not only to wait in ambush for birds and squirrels; they seem just to like to get up high to have a safe place to look down on the world. Many apartment cats feel pressured when the living room is full of visitors, and a quiet corner under a table or bed may not feel safe enough. Some will try to get on top of a closet or refrigerator, and they often knock ornaments and things over in the process. Because of cats' natural proclivity to climb up vertical objects, one way to make their environment more natural is to construct a "tree" in one corner of the living room. An actual, thick tree branch, bark and all, or a carpeted post going up to the ceiling, with two or three horizontal perches of rug or cushion, will delight many felines. The base also serves an an ideal scratching post. If the "tree" can be set up near a window, so much the better, since cats enjoy looking outdoors, especially at birds in trees or on a nearby bird feeder.

There is nothing more pathetic than an obese, understimulated pet that is simply given food and water as needed and nothing else. If we treat our pets with the indifference we might show our potted plants, they may well become as unresponsive, dull and uninteresting as vegetables. We owe it to them to cater to their basic needs and to make life as natural for them as we can. There is a condition recognized by veterinarians as the "zombie" syndrome, in which the cat, usually a purebred animal, simply lies around like an unresponsive object. This could be the final product of domestication, of breeding and rearing our pets to be quiet and docile ornaments. Ironically, some of these animals seem to be mirror images of their owners and of the owner's children, too. Some families that I have counseled about their pets fall into still different categories: the controllers and the permissive indulgers. They either completely control and inhibit all natural behavior in their pets and children or allow anything to go on, so that both pets and children are unruly and socially irresponsible. There is another category, too, that is quite prevalent: parents and pet owners who ignore their children or pets for long periods and then overstimulate them with excessive attention and overindulgence. Such inconsistency creates pets that are often confused, unpredictable, emotionally unstable and sometimes hyperactive. Children, too, often mirror such unnatural care. The healthiest and best adjusted pets and children come from those families where life is made as natural as possible for them. The parents/owners do not overindulge or discipline inconsistently. They are loving and consistent in their behavior and understanding of the needs of those in their care. Catering to the needs of a cat and endeavoring to make its life as natural as possible, just as for a child, does not mean allowing it to run wild or to rule the household. Its needs and rights, too, must be consonant with the needs and rights of others in the family and in the community, responsible parenting and responsible pet ownership being synonymous. Also, the pet must at times be protected from itself, since its natural instincts could get it into trouble. For example, an inquisitive kitten could bite through an extension cord and get electrocuted.

Remember, the more natural you can make life for your cat, the more natural your cat will be. Consequently, it will be a more alert and interesting companion to have around. By making life more natural for your cat, you will also help satisfy some of its basic needs and avoid frustrating others. In so doing, you will encourage the development of your cat's natural potential, which in turn will increase its IQ, because the more stimulating the environment is, the more your cat will respond and learn. The same also holds true for our children.

A cat's domesticated life indoors can be enriched in many ways to satisfy its basic needs. Provide your cat with a companion, ideally a kitten of the opposite sex. It is best to raise two kittens together and never to give them a taste of the outdoors, since once cats have got used to roaming outdoors, they may resent being confined at a later time.

Kittens that grow up together provide social enrichment and companionship for one another. The ideal situation is to raise one of each sex together. Their play and interaction will provide character development for each other and pleasure for the entire household.   *Photos courtesy M. W. Fox*

Two cats together are generally healthier and happier than those that live alone, though it is true that some mature cats seem to prefer only humans for company.

A pair of cats can play and sleep together, groom each other and generally take care of each other and provide for their social needs. This is especially important, considering the fact that in most homes children and adults are away all day at school and work. Cats left alone without a companion all day suffer from boredom: some become lethargic, obese, groom themselves excessively, even to the point of self-mutilation, and often become unhousebroken—a very common symptom of emotional distress.

You can also provide your cat or cats with the following: a three-foot-tall scratchpost; a tree branch or carpet-covered pole to climb with one or two carpeted shelves to lie on or boxes to hide in; expanded shelves covered with carpet material by one or more windows where they can look out and sun themselves; a screened-in balcony where the cat or cats can go out in good weather, or fit a window screen so that the window can be opened in warm weather and the cat won't fall out; an outdoor enclosure (the bigger, the better) with a tree trunk, litter box, resting boards, and one nest box so the cat can avoid unexpected showers and cold drafts. The enclosure could connect with the house via a wire-mesh enclosed runway, so that the cat or cats can go in and out whenever they desire through a flap-door cut into a back door or windowpane.

A good arrangement for cats is an enclosure at least five feet wide and twelve feet long—the longer, the better. The enclosure should be covered with chicken wire so the cat or cats cannot climb out, and half of the top should be covered with plywood or tarpaulin to provide shade and shelter. A litter box should be provided and placed inside a large box to keep rain out. The floor of the enclosure could be seeded with grass and small shrubs planted. A large tree branch could be set up inside for the cats to climb, and resting boards for sunning set up like shelves, three feet wide and two feet deep, at convenient heights along the side of the enclosure. A warm, insulated box (two feet by three feet by two feet high) with a small entry hole and a blanket inside should be provided for each cat. The cats can be put inside the enclosure during good weather.

A similar setup for cats could be arranged using a screened-in porch. They would enjoy a few pots of grass on the porch floor, and their large tree branch would be a decorator item for any sun porch.

Cats living by themselves and left without the stimulation of a feline playmate will often show symptoms of emotional distress. *Photo courtesy HSUS/France*

Kittens definitely benefit from human interaction. Those that receive adequate amounts of play from the people around them continue their playful natures into maturity.

Photo courtesy Marin Humane Society/Fowler

Many cats enjoy having a bird feeder set up outside so that they can watch the birds from the window. Some even enjoy spending time observing fish in an aquarium, which should be securely covered to protect the fish.

Cardboard boxes and large paper bags make excellent hideaways for cats and can be used for games of ambush and hide-and-seek.

Cats are especially active early in the evening, and this is a good time to play with them. Hide-and-seek games and catching "prey" such as a strip of fake fur on the end of a string will entertain them for some time, giving them beneficial exercise and strengthening their emotional bond with their human companions. Some cats, especially Siamese, learn to retrieve and enjoy having a toy mouse or a small ball of paper thrown for them to catch.

People who play with their kittens often find that their pets remain playful and attentive when they mature.

All cats should be given their own toys to play with. Many enjoy carrying around an old sock, as though it were a kitten or a dead bird.

In addition, cats enjoy a daily grooming, regular petting and even massage. And it is a good idea to grow some fresh grass or sprouted wheat in a container for them to nibble on as they desire.

With a minimum of effort and expense, a cat's home environment can be so enriched that it will thrive and its owners need not feel guilty about not allowing it freedom outdoors.

I know of several cat people who take their cats for walks on a body harness and leash, and many cats enjoy going out with the family for a drive in the car.

Because cats are so adaptable to our modern life-styles, especially to life in apartments, and since they do not need to be taken outdoors regularly as dogs must be for evacuation, their popularity is increasing. They are in many ways more convenient and less demanding of one's time and attention than dogs, who suffer much more from being left alone for extended periods and have no natural instinct or ability to learn to use a litter box.

PROVIDING A PET FOR THE PET

There's a single answer to the following very different pet troubles that owners often encounter. One of the most common troubles of cats today that owners frequently fail to recognize is that the pets are dull, disinterested in life, generally lethargic and only too often over-weight. It seems impossible that such cases could be in any way related to a cat's dying of sickness that the owner spotted too late for the veterinarian to save the animal, or a cat's grooming itself to the point of self-mutilation.

The underlying problem or common cause in these cases is lack of companionship. It is not simply human companionship, but the companionship of their own kind that many cats need. Some cats do quite well living alone all their lives with only humans, but many of these become overdependent on their owners, and this can lead to a whole set of other problems.

Many cat owners who consult me want to do what's best for their cat, and one of the best things one can do is to attempt to make the life one can provide for the cat as nearly natural as possible. Contact with its own kind means that your cat can really relate and be understood once in a while, since few of us can really "talk" cat and comprehend all our pet's intentions and actions. Being with its own kind means that your pet can develop, actualize and express its "catness"—and prob-ably enjoy it, too. Such freedom to be oneself is surely conducive to physical and psychological well-being for both human beings and an-imals alike.

The longer an animal is kept without much contact with its own kind, the more it will look to humans as a source of satisfaction for many of its basic needs. But since most of us can't behave or think like a cat, and since many owners don't really know what basic needs must be satisfied and how, some very serious problems can develop. The pet becomes frustrated or anxious and may subsequently become increas-ingly nervous and irritable. The older it gets, the more it may show avoidance, fear or aggression toward its own kind. This paradoxical reaction develops when the animal becomes overdependent on its owner and seems either incapable or unmotivated to relate with its own kind. Members of its own species are no longer a source of enjoyment. This means that with many older cats it's simply too late to attempt to normalize their lives by introducing another cat or dog into the home. The older and more set in its ways the pet is and the more humanized it is, the more likely it is to be extremely jealous of or aggressive toward any newcomer.

In my experience in veterinary practice, the healthiest-looking cats are more likely to be that way not because of their owners' knowl-edge of pet care but because they have a companion of their own kind to live with. When a kitten is young, it will engage in lots of play, especially in the early and late evening. If it doesn't have another cat to play with, it will play by itself but with less vigor and enthusiasm. Eventually, it may cease playing altogether, a probability enhanced by most owners' inability to play properly with their pet. Two cats raised together, however, may continue to play until ripe old age. My two cats, aged nine and ten respectively, engage in rough play with each other regularly several times a day. Note how cats do incredible judo and calisthenics when they are playing with each other. A daily dose of this surely helps keep them trim and supple. Lack of such stimulation, I believe, is a significant contributing factor to obesity and rapid aging, both of which are common problems in household cats.

Some owners will say that their cats never play, or stopped playing with each other once they reached maturity. While it is true that some cats don't seem to enjoy playing as adults, because they have a stoic or stand offish temperament, such animals are exceptions, I believe. Most all cats enjoy play even into old age, provided they have an owner who encourages such behavior and a companion animal for a playmate. Vigorous play stimulates the circulation, improves the heart function and muscle tone. Social play is also a creative act, stimulating the animal's imagination. In so doing, it not only alleviates the psychic sterility of a bland, unstimulating, safe and unchallenging domestic (home) environment, but also helps the animal express much of its telos or intrinsic nature—its "catness." I also feel that such animals have more vitality, enthusiasm and naturalness than a pet that has been just raised alone with people and without sufficient opportunity to be with its own kind and to develop its inner nature. The cat who has another cat or dog companion to live with is often a more satisfying and interesting pet for those people who want cats to be cats.

They are certainly more entertaining. Many of my clients and their children have hours of enjoyment watching their pets play together and sometimes joining in themselves. Playful pets can also give a home an ambience of easygoing naturalness and good humor, a factor now being recognized by some psychiatrists who are extolling the mental health benefits of owning one or more pets.

I would like to see some hard figures from some of the large veterinary hospitals to validate my suspicion that a cat is likely to live longer and have a healthier life with fewer health problems if it has a companion of its own kind to live with. Cats will frequently groom each other. This stimulation results in a decrease in the heart rate of the

There is an endless variety of activities which will amuse both feline and human. These photos show a cat having a fine time and it isn't even using any expensive props.

Photo courtesy M. W. Fox

animal being groomed, and probably of the one doing the licking, too. Such a change in heart activity is an indicator of profound physiological changes in the body that are relaxing and highly beneficial, even having potentially a healing function not unlike the laying on of hands, as I describe in *The Healing Touch*. So it is quite conceivable that two cats living together might be healthier than one living alone.

Since having a companion will make your cat more responsive and reactive, you are more likely to be able to spot it when either of your pets is getting sick. The sooner you can recognize the signs of illness, the sooner you will be able to seek veterinary help. This could mean the difference between life and death for your pet.

Vacation time is a stressful time for many cats, when they have to be put in a boarding facility or looked after at home by a neighbor. In my experience, cats fare much better when their owners are away if they have a companion animal of their own kind. Separation depression, refusal to eat and increased susceptibility to disease are all too common in pets put into board. Having two pets instead of one and making sure the two are housed together when they are left greatly reduces the chances of such undesirable consequences.

It is important to know the basic rules for introducing your cat to a companion of its own kind. First, if your cat is full-grown but not too old and set in its ways, you should not get another adult animal as its companion unless it is of the opposite sex. Cats of the same sex are more likely to fight especially when they have not been raised together. But if they are of opposite sexes, it is necessary to have the female neutered to prevent unwanted pregnancies. I would also advise neutering the male since it may show signs of sexual frustration and start to spray around the house after it has been with the neutered female any length of time. Your cat may feel that its close bond with you has been usurped by the new pet, so be sure to give the former lots of added attention; otherwise, jealousy, rivalry fights and terror could reign in your home instead of trust and affection. If you can, take the new animal companion on approval, so to speak, just in case the two don't hit it off well together. Be sure both animals have clean health records and have recently received all the "booster" vaccines your veterinarian says are needed. Cats can take a while to settle down, anywhere from a few days to three or four weeks. The resident cat may become unhousebroken, feeling its territory has been invaded. One tip is to keep the two in separate rooms for four or five days so they can get used to each other's odors before they are put face-to-face. Another trick is to put a little of the same perfume or after-shave on both cats for a few days. This helps mask the odor-of-stranger aggression-and-flight responses.

You may instead wish to get a kitten as the second pet instead of an adult cat. I have seen remarkable transformations in overweight, middle-aged, dull and "conservative" cats who take a renewed interest in life once the new young animal comes into the house. They have something, or more correctly someone, to care for, groom, play with, sleep with and even compete with for food and for their owner's attention, thus adding a variety and zest to life. But not all older cats will accept a new kitten, which may be attacked and injured if the first encounter is not carefully supervised.

Many people prefer raising a pair of kittens of around the same age together. Again, it is important to remember some of the basic rules of choosing animals of opposite sexes, which will reduce dominance fights later in life, and to be prepared to have both pets neutered as soon as they are old enough for surgery, which is generally around five to six months of age.

Some people have raised the concern to me that two animals raised together might be less responsive to their owner. This is true to some extent in some cases. My advice for those who want a more human-dependent, less cat-oriented cat is to get the second pet as a kitten when the first one is around six months of age. By this age the first pet will be well attached to its owner and will therefore satisfy the owner's need for a more human-dependent pet.

The last thing that I would endorse is any unnecessary increase in the pet population, which is a cause of considerable concern today. However, the advantages of having two pets instead of one are indeed many, and the multiple benefits greatly outweigh any possible drawbacks. All cat owners have an ethical obligation to endeavor to satisfy their pets' basic needs, and certainly providing them with the companionship of their own kind will do much to meet this obligation.

Probably the best place to find such a companion, either a kitten or young adult, would be at your local humane society adoption center, provided its felines are healthy. In this way you will not only be helping your first pet, you will be saving the life of another animal and providing it with a good home, too.

What is the best way to train a cat? Successful training is accomplished with trust, understanding and a complete lack of force. When a kitten can relate to its person as this pair seems to, a high level of training can be achieved. *Photo courtesy HSUS/Friedl*

10

Training Your Cat

IF PEOPLE ONLY KNEW more about cats, there would be less indifference and outright dislike of cats and more respect—and probably more cat owners, too! There's a widespread and erroneous myth that cats are aloof, unfriendly (except when they want something for themselves) and so independent that they simply cannot be trained. Many people prefer dogs because they believe dogs are so much easier to control and train than cats.

However, cats are highly trainable given the right understanding and attitude, and a well-trained cat makes a more satisfying companion than one raised with the erroneous notion that all cats are untrainable individualists with wills of their own.

In some ways, cats are even more trainable than dogs. For instance, puppies require days, even weeks of training to get them housebroken, while most kittens come already pretrained. They instinctively know what to do and where. Simply show your new kitten where the litter box is. Help it climb in and out, then let it explore the box alone. At half-hour intervals, take it back to the box, and when it uses the box for the first time, do not disturb the kitten. Praise it afterward and be sure that the litter box is always accessible (that no door might accidentally close, so that the kitten can't get to it) and always have the box in the same place.

With older cats that suddenly refuse to use the litter box, first have your veterinarian check the cat over for illness, and next try to pinpoint some emotional problem, such as the addition of a new kitten to the household, that might have caused the problem. Cats become unhousebroken very often as a consequence of being emotionally upset. A lapse in housetraining is one of the most common behavior problems that cat owners have to deal with. One retraining method is to confine the cat in a large cage or a small room (without carpets, such as the bathroom) with its clean litter box, food, water and a blanket to sleep on for ten to fourteen days to break the habit of "going" elsewhere in the house. Let the cat out often for petting and play but strictly supervised so that it will not evacuate anywhere in the interim.

Unpleasant odors can be used to train cats. Aversive training to stop a cat from urinating or defecating habitually in specific places outside of its litter box can be most effective, using household vinegar. Dab some vinegar on the floor where the cat has messed, pick the cat up (do not call it by name, since it will then expect some reward and won't respond later after it has been punished), and put its nose close to the vinegar. Then dab a little vinegar on its nose or gums. The cat will run away and shake its head but will not be harmed physically in any way by this treatment. Next day put a few dabs of vinegar on the same area that it habitually soils; repeat this over several days, and if the cat messes the same or a different spot again, bring it to the spot, let it sniff the vinegar, and dab a drop on its nose again. Most cats quickly develop a negative association between the smell of vinegar and the places they habitually soil, which effectively breaks the habit, and they then return to using the litter box.

Some cat owners want to train their cats to evacuate outdoors. I do not recommend this, since outdoor cats are more prone to disease and injuries from other cats and, of course, from automobiles. Start out right and train your cat early on to use a litter box so it doesn't need to go outdoors.

It may seem cruel to keep a cat indoors most of its life, but you can train your cat easily to adapt well to house or apartment life. Having a companion cat helps immensely, since they keep each other company, especially when owners are away all day.

You can train your cat to go safely outside and enjoy the sights, sounds and smells of the great outdoors and not endanger its life in the process by purchasing a harness and leash and taking it for a walk.

Many people do this, so don't feel you will be branded as a neighborhood "kook." Some cats don't like to walk too far and enjoy a ride home on your shoulders.

Scratching and raking objects in its territory are instinctive behaviors for cats. Feral cats or pets with outdoor access will leave their marks on tree trunks to establish dominion—not to sharpen their claws. *Photo courtesy M. W. Fox*

In the home, scratching behavior can be a problem when furniture and draperies become tree trunk substitutes. Kittens can and should be trained to use scratching posts and other designated items to satisfy this need. *Photo courtesy M. W. Fox*

A harness is better than a collar, since it is more comfortable, and the cat can't slip out of it if scared by a dog or a truck roaring by. Unlike walking a dog, who will follow you on the leash, you should begin by following your cat, so that it gets used to the feel of the harness. Stop whenever it wishes to sniff and explore or roll in the sun. Cats often like to just sit, say on a stone wall or park bench, and watch the world go by. Sit down with your cat and commune with nature together (but avoid areas where people take their dogs on or off the leash because your cat could be scared and even attacked).

As cats mature, nature switches on a couple of instincts that can cause problems. The first one cannot be corrected by training. This is spraying. Male kittens, as they mature, start to spray objects in the house with their urine. This is such a deeply ingrained instinct to mark their territories that no amount of training will inhibit them. The solution is to prevent it by having the kitten neutered around four or five months of age.

The second instinctual behavior is clawing upholstery. Cats do this out of sheer excitement and to mark their territories. Outdoor cats and wildcats rake tree trunks with their sharp claws, not simply to sharpen their claws, as many people believe, but to leave an impressive mark to display their territorial occupancy to rival cats. This instinctive behavior will quickly ruin furniture and drapes—and even wallpaper. So as soon as kitty begins to show this behavior, buy one or two scratching posts, upright posts, say six inches wide and three or four feet high, covered in carpet. Put the cat up against the post and place its forepaws against the carpeting and rake its paws up and down. Claw the post yourself—cats *are* copycats. A piece of carpeting tacked against a wall or on the side of the sofa will also be attractive. I prefer training a cat this way to having its claws surgically removed. Not all cats will stand having their claws snipped short every few weeks, but for those who refuse to use a scratching post and will tolerate having their nails clipped, I opt for that alternative over surgery any time.

Kittens, especially those with an excitable, feisty disposition, need to be trained not to claw and bite too hard when they are playing. It distresses me how many pet owners are quick to give up on such fine kittens, believing them to be too wild and untrustworthy. All they need is some disciplinary training. Mother cat wouldn't allow such liberties, and we can copy her behavior in correcting a rough-playing kitten in the language that kittens best understand: cat-talk. As soon as kitty bites or claws too hard, hiss loudly *"Tssst,"* shout "No" and then tap the kitten on the nose just as mother cat will give a quick, light swipe with her paw. A nose-swipe like this startles the kitten and gets the message across effectively without causing the slightest physical injury. One never needs to actually hit a cat to discipline it.

I would go so far as to encourage a kitten to play roughly so that this *"Tsst*—no—nose-tap" training can be instigated. The kitten will quickly learn to respect you as the dominant parent figure and the *"Tsst*—no" can be used in other situations later in life to stop other undesirable behaviors. The *"Tsst*—no—nose-tap" is actually a conditioning procedure from which the kitten quickly learns that soon after *"Tsst*—no" (or a hiss from mother), its nose will be swiped at. Consequently, as soon as it hears mother cat hiss or you hiss *"Tsst*—no," it will cease and desist whatever it is doing in anticipation of having its nose swiped. So once it is conditioned, nose-swiping need not be used anymore. Occasionally perhaps, just to strengthen the conditioning. And now with my cats, whenever they jump onto the kitchen counter, refrigerator, dinner table, get under my feet or decide to chew on a potted plant, a loud *"Tsst*—no" effectively prevents them from repeating such undesirable pranks.

An alternative conditioning for such aversive training is to precede a squirt of water with a plastic plant mister with the *"Tsst*—no" signal to train your cat not to eat your plants or leap onto the dining table. After a few squirts you can lay down the mister, since the *"Tsst*—no" will suffice, except for the duller or more strong-willed felines who will need an occasional "reminder" squirt to keep them in line.

Not all training of cats entails such schoolmarm discipline and rigidity. Again, let's look at how mother cats train their kittens and learn some more "cat-talk" that we can use with our own cats.

As mother cat will hiss and spit to discipline her kittens, she will also give other specific sounds that either call them to her or maintain close, friendly contact. Since mother cats don't call their kittens by name but use a specific musical meow-call, the naming of cats is less important than the tone of voice you use. The more like a meow-call it is, the more readily kitty will learn to come when called. So whatever the cat's name is—Gandalph, Darwin or Sam, Fluffy or Prudence—train it first to come to you when its name is called in a friendly, musical tone, just for food.

Have some special tidbits to reward the cat for coming to you from anywhere in the house whenever it hears its name given in the mother-is-calling tone of voice.

This kind of voice-signal training is extremely valuable when the cat slips outdoors and you can't find it. If people would train their cats to come when called, the cats would meow back, as though to say "Yes, I'm coming" (just as kittens do when responding to their mother's call), and fewer cats would be lost forever. I have seen lost cats crouching in fear under a bush or parked car as their owners call for them in vain. Properly trained, such cats would respond immediately.

You can also use a different tone of voice to discipline your cat, giving a hissing or growling inflection as you speak its name. Or you can repeat its name in a quiet continuous purring tone, just like mother cat purrs to her kittens while they are nursing or being groomed. This will help in training your cat to be still while being brushed or having its claws trimmed and while being examined by the veterinarian.

Voice control in these three basic modes—to seek contact, to discipline or break contact, and to maintain close contact—are essential ingredients of cat training, which, since the sounds so closely mimic the natural vocal repertoire of cats, will be quickly learned, especially when training starts early in kittenhood.

Once the basic voice-control signals have been learned, you can use the appropriate disciplinary tone to train your cat to sit and stay. First make eye contact and place your hands over the cat as though to hold it down and simply say "stay" or "sit" in a firm voice. Then back away a few paces, keeping your eyes fixed on the cat and arms extended, palms down as though you were still holding the cat down, and keep repeating the word *stay* or *sit*. Reward the cat by petting it and talking to it gently after it has remained still for about fifteen seconds. Extend this time period for up to a minute as training progresses.

If the cat gets up or tries to run away, gently pick it up and return it to the chair or area of floor where you are training it. A leash and harness will facilitate control. Patient repetitions, no more than ten trials per day (more would be stressful overtraining), should suffice.

Many cats can also be trained to fetch and retrieve objects. Encourage your cat to chase a ball of wool or suitable cat toy such as a rubber squeak-toy or furry "mouse" stuffed with catnip. Have a long leash on the cat. After the cat has caught the toy and played with it briefly, and while it is still in the cat's jaws, call the cat to you and gently encourage it to come by pulling on the leash. Praise the cat when it reaches you, wait a couple of seconds and then take the toy away from it and throw the toy again for the cat to chase. The cat will

Cats enjoy their comfort, and a secure feline considers comfort its right. When you train, do so with an awareness of the comfort factor and the cat's great sense of self-esteem.

Photo courtesy M. W. Fox

quickly learn (if it enjoys this game) that it is more rewarding to chase and catch the toy and bring it to you to throw again than it is to go off and play alone with the toy. Some cats are spontaneous retrievers and actually train their owners to keep throwing their toys around for them to chase.

Cats can also be trained to perform simple and entertaining tricks such as catch, sit-up-and-beg, and jumping either through the air or into your lap. Teach catch by having your pet close by and throwing a ball of tinfoil or wool just over its head. You can start with the ball on a string, swinging it above the cat so that it can bat at it and catch it. The cat will rear up on its hind legs, and in this position you can, if you wish, teach it to beg for food by waving a piece of fish or other delicacy (asparagus works fine for my Abyssinian) over its head. Once you have the cat's eyes fixed on your hand, move your hand in an upward and then downward arc to make the cat jump. Soon it will stand and beg or jump without food being in your hand, since it becomes conditioned to the position of your hand and your hand movements. But begging and jumping will quickly wane if you don't give the occasional food reward. Many people like to train their pets to perform such tricks to impress friends, but I'm unimpressed: it is a form of teasing, since the cat is anticipating a reward of food each time. My cats readily jump onto my lap when I call them as I pat and stroke my hands on my knees, the signal being obvious: if you jump onto my lap, you will get stroked!

Cats train us in a number of ways—what kinds of food they prefer, when they like best to be fed, where they want to sleep and so on. We should be alert to the fact that since cats can and do manipulate and condition us to satisfy a variety of wants, they may, in the process, either become spoiled overlords of the house or eat things that are not good for their health. Some cats will go on hunger strikes, refusing to eat anything except canned tuna or fried chicken. Such food, as a regular diet, is inadequate and unbalanced. Don't let your cat do a number on you!

A cat will sometimes seem to act klutzy or seem to forget all its training and good manners, knocking an ornament over or urinating on the floor. Don't dismiss such aberrant behavior as stupidity or forgetfulness of training. It could be trying to communicate with you. Some cats have urinated in obvious places because they have a serious and painful bladder infection and need to see the vet, or want to go outdoors to eat a little fresh grass. Cats have knocked ornaments over and generally acted wild not just to get an early breakfast but because there's a fire or flood in the basement. So don't scold a well-trained cat that suddenly seems to behave foolishly—it may well have something to teach you.

Training cats, like training children, is for their own good as much as for the convenience of owners and parents. Furthermore, an educated, well-adjusted, well-mannered and responsive cat is a delight to live with. A little time spent training kitty will also lead to a closer attunement and reciprocal understanding and respect.

POSTSCRIPT

Cat-Litter Precautions

Sensible pet owners know that if they want to change the kind of food they are feeding to their companion animals, they should do so gradually, simply by increasing the amount of the new food that is mixed in with the cat's regular food, which is proportionately reduced in quantity each day. Otherwise, the cat may turn its nose up at the new food. The same holds true when you want to change your brand of litter for various reasons. It is best to stick to the same brand because once a cat gets used to a particular kind of litter, it may refuse to enter its litter box if a different brand with a different texture and odor is used. My advice is, if you must change the cat's litter—especially if the brand you're using is extremely dusty, which could be a hazard to your cat's health as well as causing tracking of litter outside the box—you should switch brands gradually. Simply put an increasing proportion of the new brand of litter into the box each time you clean it out. I find that adding about one quarter of the new litter each time generally works best. Cats who develop a new-litter phobia because the new litter has not been introduced gradually can suffer from urine retention, constipation, and may even become unhousebroken and start evacuating in various parts of the house.

Note: Some cats are extremely fastidious and like a clean box. This means that all stools should be removed once a day. Otherwise, they may well refuse to use the litter box. Many cats prefer to have the litter box in a safe, quiet corner and may refuse to use it if it is relatively exposed in a place where there is frequent household activity going on.

It is a hopeful commentary on humankind and contemporary society that more attention is being paid to animal rights. All around us we see greater attempts to understand the animals in our world and improve the overall condition of their lives. *Photo courtesy M. W. Fox*

11

The Four Rs of Cat Rights, Health and Owner Responsibility

WISDOM'S "FOUR PILLARS" of veterinary preventive medicine and health-care maintenance are as follows: *right understanding, right environment, right breeding, and right feeding.* These are the basic rights of all animals under our dominion.

These four principles are complementary and synergistic. A weakness or deficiency in any one of these may be compensated to some degree by the others. The probability of disease increases in proportion to the degree of weakness or deficiency in any one of these four pillars, and also increases with a lack of synergy or compensatory support between them. For example, without right understanding, the problems of a cat afflicted with an unstable temperament (which could be a product of wrong breeding) can be exacerbated. Likewise, with wrong feeding, the health problems of animals with a genetic predisposition to metabolic or endocrine disorders can be aggravated.

Such are the dynamics of these four interrelated and interdependent principles. *Right understanding* entails knowing and satisfying an animal's behavioral and emotional requirements. This includes proper socialization and rearing to establish a close emotional bond between the person and the animal, where the animal is neither overindulged, neglected, nor subjected to excessive control or physical or psychological abuse.

Right environment entails providing the animal with conditions conducive to the expression and satisfaction of its basic physical and psychological needs in such a way that its adaptability to domesticity is not jeopardized. Opportunities for exercise, play, affiliation with other animals (such as keeping two cats instead of one), are all part of providing the right environment for your cat. The animal should also be provided with various toys and an enclosed outdoor pen to enhance that environment.

Right breeding entails careful attention to the elimination of inevitable disorders that arise as a result of inbreeding and selection for abnormal traits in purebred cats. An alternative is to promote the ownership of cats, mongrel or mixed-breed, with a significantly lower incidence of such anomalies, which can cause unnecessary suffering and increased susceptibility to disease.

Right feeding entails providing animals with a wholesome, balanced diet. This will vary according to the breed, size, temperament and age of the animal, as it does also when the animal is growing, pregnant or convalescent.

Greater emphasis needs to be placed upon these four principles in the education of veterinary students and in the education of cat owners and breeders. Since these animals provide people with the many benefits of their companionship and unconditional love, we surely owe them no less: right understanding, right environment, right breeding, and right feeding. These principles may be regarded as the basic rights of companion animals, and their recognition will contribute significantly to the prevention of unnecessary suffering and to their overall well-being.

LEAVE IT TO NATURE?

There is a widespread attitude, if not a superstition, that when an animal is sick, "Nature will take care of things." It is difficult to convince some pet owners that nature takes care of nothing. Animals in the wild take care of themselves (as by eating medicinal herbs, fasting, withdrawing and resting) and of each other (sick or injured wolves are often fed, groomed and protected by pack-mates).

It is mutually advantageous to respect animal rights for humans and the animals in their environment. The confidence shown by this well cared for Abyssinian is a reflection of its excellent upbringing and regular care. *Photo courtesy M. W. Fox*

Left to their own devices, cats will engage in such activities as eating grass for the sake of their general well-being. It is wrong, however, to assume that a sick cat will always be restored to health by natural means without human help. *Photo courtesy M. W. Fox*

117

To let nature care for sick or injured pets is the height of irresponsibility, since the house pet is neither natural nor a part of nature.

Leaving it to nature can also be a kind of denial. People don't want to have to worry about a sick pet or about vet's bills, so it's more comfortable and convenient to leave a sick animal alone.

While veterinarians don't agree with all my views on animal rights, they are unanimous in their agreement that the leave-it-to-nature attitude is a widespread problem in the pet-keeping public. They recognize that animals have rights, or at least one right, and that is to receive proper veterinary attention when they are sick or injured. (No, they aren't simply rooting for more business.)

COAT AND SKIN CARE: BEAUTY AND THE BEAST

Not too long ago I saw a beautiful coyote loping across the desert out West in the Sierras. It was early morning and the low light shone off the animal's lustrous coat as though the coyote had been wrought of burnished bronze and gold. Wild animals do radiate health and vitality, and indeed it is the quality of the animal's fur or "pelage" that clearly reflects its overall physical well-being. This may be one of the reasons people like to have their pet's pelage looking neat and lustrous. No one enjoys looking at a disheveled, unkempt animal, or person for that matter. Outward appearances do give an indication of inner physical and also psychological well-being. As a sick cat will not groom itself and will quickly become an unesthetic, pathetic looking creature with no "bloom" to its coat, likewise a depressed person will often have little interest in his or her personal appearance. We all know, at least unconsciously, that our own outward appearance affects not only how others feel toward us but also how we feel about ourselves. Well-groomed and neatly turned out, we get the attention and affirmation of others, and our self-esteem and self-confidence rise, and likewise when we dress up our children and people compliment us on their appearance. So, too, when we do a fine job grooming our cat, people notice and we get complimented. Our self-esteem rises. We also enjoy the presence of a handsome or beautiful animal bubbling over, at least on the surface, with health and vitality.

But does looking good do anything for the pet's sense of self-esteem or overall well-being? Some people have told me vehemently that a good grooming or thorough "beauty treatment" does wonders for a cat's self-esteem. Others say that that's all anthropomorphic nonsense, and that animals are incapable of having such human feelings. So who's right?

Making your cat look beautiful will make it feel good, and I will prove that in a moment with some clear and concrete scientific evidence. Whether or not a good grooming or beauty treatment improves your pet's self-esteem or self-image is something else. I, for one, do not doubt that animals have a sense of self, but grooming per se will not, I believe, give a boost to your pet's ego. Being complimented on having a beautiful pet brimming over with vitality is certainly a boost to the owner's ego. And I am convinced that a pet that does look beautiful enjoys the extra attention that its good looks evoke from people. This isn't being anthropomorphic. Observe how a well-preened cat waits until all the guests are settled before she makes her grand entry into the room.

It probably matters little among cats whether they are looking good or not—how they smell is probably just as or even more important than how they look. But living in a humanized world as domesticated animals, cats learn that when they are looking good, they receive more human attention. Few can resist wanting to touch a freshly groomed and puffed-out Persian or a sleek and shiny Burmese.

The scientific evidence for the beneficial effects of grooming I have described in detail in my book entitled *The Healing Touch*. Briefly, grooming, like massage, stimulates the circulation and the hair follicles of the skin, helping remove loose and dead hair and scales of skin (which are shed and replaced cyclically). Grooming adds sheen to the coat by releasing and spreading natural oils from the oil glands in the hair follicles or ''roots.'' Since grooming stimulates the blood and lymphatic circulations, as does massage, it is extremely beneficial for aged and convalescent animals. Such stimulation may not only be invigorating, it may also promote the growth of a luxuriant coat. The act of grooming, like that of petting, results in very deep relaxation and a dramatic slowing of the *animal's* heart rate rhythm, which, as I describe in my massage book, is not only a pleasurable sensation for the animal, it will also help a young or sick animal digest food better and will certainly help animals cope better with and even overcome certain physical and emotional stresses and sickness in general. My cats all come running when they see me with their brushes, ready to give them a thorough grooming, a clear sign of how much they enjoy it.

Skeptics might say pets don't need beauty treatment, since, after all, their counterparts in the wild don't have anyone to groom them and keep them looking in top condition. However, in the wild, the animals and living conditions are completely different from our environment. We keep pets that are genetically very different from wild animals, and wild animals don't need any help in keeping healthy and looking good. Having to hunt actively every day keeps wild cats supple and trim, and as they are in harmony with the seasons, their coats are thick and lustrous in the winter and light and airy in the summer. Keeping our pets indoors can disrupt the seasonal cycle of coat shedding and re-growth. And we have changed cats genetically, so that some have abnormally long fur, never shed or have thick underfur year round. Thus, as a consequence of indoor living and genetic changes, most cats need some help in keeping their coats in good condition. Also, they do not hunt and eat their own natural foods but instead enjoy an easy life of being given processed commercial foods and table scraps. Diet will influence how well your pet looks, and nutrition certainly affects how lustrous your cat's coat is. More of this later. There are some additional advantages to consider from giving your cat a daily grooming and a beauty treatment (which I will detail shortly) every few weeks.

Cats with long coats that tangle easily require daily grooming; otherwise, tangled balls of fur will quickly develop beneath the ears, on the "feathers" or long fur on the legs, especially between the hind legs and along the tail and belly. In veterinary practice I had on many occasions to anesthetize animals in order to remove large mats and balls of fur, which can cause considerable discomfort, stop the skin from being properly aerated and lead to skin infections beneath the mats. Owners complained that their pets would not tolerate being groomed, or that it was too much trouble. However, having to tran-quilize or anesthetize a cat every time it has to be thoroughly groomed is a health risk. Repeated treatments can result in liver damage and increasing intolerance to anesthesia. But there is nothing else to do if an animal is frightened or acts aggressively when one attempts to groom it, snip away mats of fur and trim its feathers. The solution is simple: start out right from the beginning with a kitten, especially with a long-haired variety, and teach it to submit quietly to regular groom-ing sessions. The earlier in life a kitten gets used to being groomed and handled, the more willing it will be to be groomed, and the easier it will be to handle later in life. Cats often like to flop over onto one side to be groomed, so don't insist that your cat sit up or stand. Work on one side and then just grab hold of the front and hind legs and roll it over to do the other side. Some cats don't like to be groomed near the

Longhaired cats were developed by man and so require human assistance with their grooming. Without regular, thorough attention, the average longhaired cat can become matted, filthy and an easy victim for skin disorders and parasites. *Photo courtesy Mary Pyles*

Notwithstanding the advantage of easy grooming care presented by shorthaired cats, regular grooming for them will benefit coat and skin health and reduce the amount of ingested hair normally swallowed during self-grooming. *Photo courtesy Mary Pyles*

121

base of the tail or along the abdomen, so avoid much contact with these hypersensitive areas in sensitive cats; otherwise you may be scratched and, furthermore, make the grooming session unpleasant for the animal.

Another major advantage of grooming cats on a regular basis is that the more loose fur you remove, the less there will be for the cat to swallow when it grooms itself. Consequently, fewer fur balls are likely to build up in the stomach, and this will reduce the frequency of your cat's vomiting food and fur on your rugs and furniture.

Finally, when you are grooming your pet and while giving it a beauty treatment, you will be able to feel and see any abnormalities, such as a swelling or red or scaly area of skin that may require veterinary treatment. Owners who groom their pets on a regular basis are more likely to spot when their pet may have something wrong with it, and when a problem is identified early, veterinary treatment is often more effective than if it has been allowed to fulminate unnoticed for weeks. This could mean the difference between life and death.

Some groomers like a variety of different combs and brushes. For cats I prefer a fine stiff-bristle hairbrush. To begin grooming, I call the animal, get it settled and then stroke it reassuringly around the head. I run my fingers down the animal's back quite hard, several times, to loosen up any dead fur, and then push my fingers into the pile of the fur and work upward from the animal's tail up to its head.

If the atmosphere is very dry and "staticky," especially in winter, I moisten my fingers (and the brush) first and only groom the animal on a wool rug or cotton towel, to reduce the chance of giving the animal a painful electrical shock. Vigorous brushing and this finger-stroking grooming-massage, especially with the animal on a nylon carpet in winter, will charge the animal up and give it repeated electrical shocks.

After loosening the fur with the fingers, I groom the cat in slow, deep and long strokes from head to tail. If the animal is shedding heavily, I make short strokes as I work down the back, removing the fur as needed from between the bristles of the brush. For long-haired cats, I twist the brush outward, away from the body, and brush through the fur, also along the tail, so as to fluff out the fur and stop it from getting tangled. I moisten the brush slightly if the fur gets "staticky," because it is more likely to get snarled then.

Don't brush too hard or vigorously, or else you may actually cause bald spots in some cats. Many cats have "loose" coats that shed fur continuously, and they will look unsightly if groomed too much in one spot! After a thorough brushing along the back, tail, down the legs and along the sides and belly, always following the line of the fur, I stand the cat up and give it a few gentle strokes with my hands to give its coat a final smoothing polish.

Always apologize if you accidentally knock some bony protuberance such as a knee or shoulder with the edge of the brush and give the animal a reassuring stroke as well. Aggregations of fur on the floor after you finish can be vacuumed up or, if very clingy with static, wiped up easily with a moist sponge.

Poor diet certainly influences the quality of an animal's pelage and its overall appearance. If your cat has a dry, lackluster coat, it probably needs more vegetable oil in its food, such as linseed, sunflower or safflower. Give one teaspoon per fifteen pounds body weight daily, and once the coat looks good, reduce that to one half teaspoon per fifteen pounds body weight. Dry pet foods tend to be low in polyunsaturated fats and high in carbohydrates, so in addition to giving vegetable oil to boost the fat in the diet, try feeding a high-protein moist canned food. The addition of a few drops of vitamins A, D and E, and a pinch of powdered kelp (seaweed) and brewer's or nutritional yeast (working up to a quarter teaspoon per fifteen pounds body weight daily) is a kind of internal beauty treatment that will pay dividends. A chronically poor-looking coat needs veterinary attention, since diet alone won't rectify a hormonal imbalance, leukemia or worm infestation, three of a number of factors that may make a cat look run-down.

Cutting a cat's fur is sometimes necessary for the animal's well-being. And sometimes it isn't. Owners are often confused over this, which at first sniff seems like a trivial issue. For the cat, though, it often isn't.

Take, for instance, the owner who wrote to me for advice over suing her vet, who had given her cat a patchy haircut. The animal had "holes" in its coat and was so unsightly, the woman was embarrassed to have anyone see her cat until the hair had grown back. But in this case the animal was suffering from a mild skin infection, and the owner thought the problem could have been cured with an injection and some ointment rubbed into the affected areas.

However, it is bad practice not to remove the hair over quite a large area to stop skin infections from spreading.* Furthermore, the

* Note: Animals should be discouraged from licking sores only if they are doing so excessively and delaying the healing process, because saliva contains its own healing agents, which normally accelerate wound repair and recovery from infection.

hair that lies over a "hot spot" or other skin disorder provides the ideal breeding ground for bacteria, especially secondary staph infections, which produce smelly pus and delay the healing of wounds, sores, hot spots and other skin maladies. So the more hair that is removed, the better and sooner will the skin heal.

It is advisable to trim the "feathers"—the long hair on the hind legs of long-haired pets, especially those that are overweight and cannot clean themselves easily. This will prevent the fur from matting into balls, which can cause some discomfort. If these balls get soaked in urine, they can be a source of infection as well as an unpleasant odor. It is surprising how many people will let an animal suffer for quasi-esthetic reasons, rather than getting the scissors out and giving it a good trim—or letting a groomer do the job.

BATHING YOUR CAT*

Bathing one's cat, especially for the first time, can be a comedy of errors. Cats especially are "aquaphobic" and may run up your body as though you were a tree, leaving claw-marks of remembrance. They can become hysterical and defecate, urinate, attempt to escape, even bite, or act as though they are being punished. So one must be prepared, since few pets take to having a bath for the first time. And subsequent times can be even more difficult if the first experience is unduly traumatic.

First, does the animal really need a bath? Cats always "bathe" themselves by licking and grooming themselves thoroughly. Healthy cats never need a bath. A daily grooming with a bristle brush is sufficient. But baths are sometimes needed for various skin ailments. If you aren't a competent cat handler, I would advise you have your cat bathed professionally at the veterinarian's or by a pet groomer if the animal doctor says that your cat needs to be bathed.

Older cats, and some "lazy" individuals, especially those with very thick and long coats, don't groom themselves as best they might. Obese cats can have great difficulty, especially reaching their hindquarters, and fatty skin folds can become soaked in urine and produce awful odors. In addition to a daily brushing, a dry shampoo with baking

* Bathing a cat once a month significantly reduces dander and helps people allergic to cats live in more comfort with them.

124

soda (best done outdoors) every couple of weeks will help keep such cats smelling sweeter and no doubt feeling better. Surgery may be needed to remove large skin folds in obese cats, along with a reducing diet.

Cats that get outdoors often come home with tar in their coats, or grease and oil from crawling under cars. If they try to clean themselves and swallow such material, they can get sick, and if they aren't cleaned, their coats will become matted and their skin irritated and infected. Tar can be removed with linseed oil. Rub the oil into the fur and then remove it with a dry cloth. Gum comes out easily after you rub some cold cream or peanut butter into the gummy fur. You can absorb grease in your pet's coat with cornstarch or talcum, removing it with a fine comb prior to bathing. Burrs can be removed with oil or crushed with pliers and then brushed out.

To clean a cat after an emergency—such as the animal's getting covered in tar, antifreeze, or garden weed or insect killer—it is best to give it a sponge bath. Place the cat in the empty bathtub or sink, holding it by the scruff of the neck. Then sponge it all over with warm water from a bowl. The water should be slightly above body temperature (100 degrees F.). Once the coat is wet, rub in a mild baby shampoo and then rinse it out either with warm water from a watering can or kitchen sink spray-hose. Dry the cat thoroughly in a soft towel and finish off with a hair dryer if the animal is not too frightened for you to use one.

Never try to wash the cat's face while bathing it, since this could cause it to panic. Cats will also panic if they get water in their eyes and ears. A plug of cotton in each ear will help keep water out. If the face needs to be washed, do this separately. A little Vaseline around the eyelids will help keep water and shampoo out of the animal's eyes. If the ears or eyes need cleaning, see your veterinarian, because the animal could have a chronic infection that might be aggravated by bathing.

SYNOPSIS: BENEFITS OF REGULAR GROOMING

A regular grooming and attention to the animal's coat and skin is part of the responsible custodianship of our four-legged companions. Indeed it is one of their "rights," and the benefits are many. I documented in my book on massage therapy entitled *The Healing Touch*

In social groups such as this one, grooming is one of the activities cats perform for each other. When there is only one cat in a household, grooming sessions are essential for the cat's happiness as well as its coat.   *Photo courtesy Myanmar Birmans/David Hovell*

that the social activity of grooming has a bonding function, as between mates and parents and offspring. And grooming causes a marked decrease in heart rate, part of the "relaxation response."

Social animals, such as cats, often groom each other on a regular basis, so it is quite natural for companion animals to enjoy being properly and regularly brushed and combed and occasionally bathed. It *is* unnatural, especially for a cat deprived of the companionship of its own kind, never to be groomed by its human companions. This is one of the reasons why two cats living in the same house together are often happier and healthier than those who live just with humans.

Grooming, more so than occasional petting, will evoke the physiologically beneficial relaxation response. This will not only bond your animal companion closer to you, it is also a wonderful way of communicating and of expressing devotion—a kind of communion, if you wish. The beneficial physiological and emotional stimulation grooming provides will also improve the animal's health and immune system because the relaxation response helps rest the overworked adrenal-gland stress-response system. Many overweight and lethargic cats are understressed. For them, a vigorous grooming is a "pickup," and is wisely followed by games, exercise and any activities that will arouse the animal to play and explore.

The environment (especially artificial heat and light) disturbs cats' natural, seasonal, coat-shedding and hair-growth cycles and can even affect the pigment of the hairs in some animals. Thus many indoor cats need constant grooming, because they are shedding and growing hair constantly. This could be physically stressful to some cats, aside from the fact that it makes them prone to gastrointestinal problems, due to fur balls forming from swallowed fur licked out of the coat when they groom themselves.

Genetically altered coats, ranging so widely as they do, need special attention and often professional help. Many cat breeds could never exist in the wild for long with the kinds of coats that we have given them. This makes it even more our responsibility and our moral duty to care for their coats and grooming needs. It is a cruelty of neglect to not properly provide for an animal's grooming requirements.

Genetically deformed animals, and the obese, aged, arthritic and chronically sick ones who cannot groom themselves properly also need special attention and more frequent bathing.

Children and cats can be a good mix when parents exercise reasonable supervision in the home and teach children early that the family cat has rights and feelings.

Photo courtesy Bonnie Smith

12

New Babies
and Young Children:
Nipping Problems
Before They Start

ONE OF THE MOST serious concerns for many parents is how their cat will accept a new baby. The best way to cope with these concerns is to be informed, and I will spell out what problems you may encounter and what you can do to nip them in the bud before the baby or someone in the family gets nipped or worse.

The last thing that I want to do is to remove all anxiety. Parents must be constantly vigilant and never feel so secure with their pet or with their knowledge that no accidents could happen.

On the other hand, I don't want to scare prospective parents to the point of getting rid of their cat. Some doctors have been very irresponsible in this regard, advising pregnant mothers as a matter of routine health care to get rid of their cats to eliminate the possibility of prenatal parasitic infection called toxoplasmosis. Others have ruled out keeping a cat because it can harbor other parasites (worms) that can infest children. However, any health expert will tell you that infants are at greater risk when they are around humans than around animals, because most diseases are carried and transmitted only by humans.

This brings us to the first rule of preparing for the arrival of a new baby. Consult your veterinarian, who will run tests to make sure that your cat is healthy and safe before the baby's arrival.

Second, many cats sense that something is going on in the home, especially during the last trimester of pregnancy, and this can make them more anxious and lead to some problems. A mother's apprehension over or preoccupation with pregnancy can affect the cat because the usual routines change and the animal receives less attention. Some cats withdraw, or become more solicitous, and this can make things worse because they may be punished for being a nuisance.

A pregnant woman can still pet, groom and talk to her companion cat, and given some indulgence, most cats adjust well. The reduced activity of the expectant mother and the increased focus of attention on her by the husband can make a very dependent cat feel rejected. This isn't anthropomorphic speculation. Many couples today have raised their cats to be very dependent, in some ways using them emotionally as child substitutes. I have had letters from couples who wonder why their cats have become less affectionate and even become unhousebroken after the arrival of a baby. So the second rule is to be sure to put time in with your cat so that it does not feel neglected or rejected.

A few weeks before the baby is due to arrive, it is advantageous to buy a doll, swaddle it and pretend to nurse and cuddle it. Let your cat see what's going on and become habituated to your new behavioral repertoire. A tape recording of a baby crying, played occasionally from within the swaddled doll, will also help your cat more easily accommodate the new baby, since it will be less alarmed by the novelty of its presence and your focusing attention on the infant.

When the baby comes home, even more attention must be showered on the animal to allay jealousy. Cats do experience and express jealousy, what amounts to sibling rivalry, which is more intense the more dependent the cat is to begin with.

This can be expressed in a variety of ways. Some cats will become unhousebroken, even urinating or spraying on the baby's bedding and laundry. One cat began using its owners' bed as a litter box! Others become excessively solicitous, especially when the baby is being fed or having its diaper changed, or will follow the owner around

The arrival of a new baby in the home may trigger behavior changes in the family cat. Normal habits may be carried out to an exaggerated degree and undesirable patterns may form. In most cases, problems can be avoided by giving the cat enough understanding and attention so it is not pushed into the background.   *Photo courtesy M. W. Fox*

Considering the cat's normally territorial nature, infants and toddlers should be discouraged from invading a cat's space. It is common for injuries to result when a resting cat, such as this, is disturbed by a young child.   *Photo courtesy M. W. Fox*

the house and resent being left alone. When the cat becomes a nuisance in this way, it is only too easy to forget why it is behaving as it does and to punish it. This will make things much worse. More petting, praise and reassurance are indicated.

Some cats will hiss and run away when they see the baby, exhibiting fear over the intrusion of a new being in the home. And because their territory has been invaded, some cats will start to spray or defecate, a behavior motivated by insecurity and probably intended to mark their home base with their own familiar scent so that they feel more secure. Consult your veterinarian. A mild tranquilizer may help, but first the animal doctor must rule out the possibility of some physical problem such as an inflammation of the urinary bladder. Emotional distress can bring on such health problems.

The third rule is to help satisfy the cat's curiosity and allay its fears by introducing it to the baby. Hold the baby protectively in your arms and sit down so the cat can come close and look and sniff. Talk in a quiet, calm and reassuring voice. The animal may be alarmed when the baby cries, but should quickly habituate to this. I know of some cats who will immediately come to a parent when the baby is crying, as though to tell him or her to hurry up and take care of things. In fact, the cry of a baby is not unlike the distressed mew of a kitten, and most cats seem to react with understanding.

The sudden, jerky movements of a baby's limbs can sometimes alarm a cat, or become an attractive stimulus for the animal to paw at or bite at playfully. Such reactions should be inhibited by a word of discipline, and they are one reason why the final rule must always be followed. This is to *never* leave the baby alone and unsupervised with the animal.

As the baby matures and is able to reach out and grasp things, such as the cat's tail, this rule becomes even more imperative. The cat has to be protected from the child. Young infants have a tendency to grab and hold very hard and have no understanding that they are causing pain. Any mother with long hair knows this.

Most cats are extremely understanding, patient and long-suffering with infants, but even with a wholly reliable cat, accidents and tragedy can happen. Often when a cat has had enough contact with a toddler, it will get up and move away. The toddler should be discouraged from crawling or walking after the animal, and a safe "off-limits" place should be set aside for the pet where it can rest in peace. The place where the cat eats and drinks should also be off-limits and made childproof with a gate in the doorway, so that when the cat is being fed, it can eat without any interference. Some cats when they are eating will act more aggressively, and a warning hiss could mean an accidental clawing. It is natural for a cat to hiss protectively over its food when it feels threatened by the presence of a small child or adult. Children should be advised, once they are old enough to understand, to behave quietly around animals, especially those animals whom they do not know and who do not know them.

The myth that a cat will climb into a baby's crib to suck out its breath and kill it probably arose from the sudden crib death syndrome, and cats got the blame. But since crib death occurs in families where there are no cats, this myth can be put to rest. However, there was a case in Europe where a large cat cuddling next to a baby actually smothered it to death.

Growing up with a responsive pet provides the child with companionship and security; a sense of being important and loved when greeted by the animal; a sense of empathy when the pet is sick and when the pet responds to the child when the child is sick or unhappy; a sense of competence and heightened self-esteem when a loving animal obeys and submits. And ultimately through such love from another being, a child will come to be humane, compassionate and respectful of all life. But if a child is wholly deprived of all animals, this eventuality is most unlikely.

There was a time when cats were ruthlessly tortured and slaughtered because of the fears of a superstitious world. This persecution was widespread throughout Europe, and the decimation of the cat population helped spread plague over a vast area. In our enlightened society, we no longer invest cats with dark powers, but we still have a long way to go in matters of animal awareness and animal rights. *Photo courtesy M. W. Fox*

13

Conclusions: Animal Awareness and Animal Rights

HOW OFTEN have I heard people pontificate that since animals can't talk, they obviously can't think or reason. The belief that animals act automatically, their minds and actions being controlled by reflexlike instincts, is widespread.

Such thinking may be derived from the old Cartesian philosophy that animals are unfeeling machines. Instincts are seen as fixed programs that control and regulate the animal's life, much as a program fed into a computer "brain" controls the workings of some industrial machinery.

While parts of our brains function not unlike a complex computer, we can "program" ourselves. We can be rational, objective, change our minds, develop new programs on the basis of experience (learning) and so on. We have insight, reasoning and intuition above and beyond the hard logic of computer functions. But to accept that some animals also possess such faculties is a heresy in many circles. A common retort to such claims is that one is being anthropomorphic—making animals into "little people" and projecting onto them human attributes that they can't possibly possess because they are *only animals*, and animal means irrational, instinct-controlled automaton. Some take this one step further and suggest that they can't be intelligent because they not only have no language, they don't even have immortal souls like humans.

Blind faith in such enculturated ways of thinking prevents people from really seeing, understanding and appreciating the nature of animals. They cannot understand that instinct is the wisdom of nature genetically encoded in the animal. Instinct is something to stand in awe and wonderment of, not something to demean as subhuman. We, as animals, possess many instinctual actions, such as the nursing movements of a baby, infantile "babbling," smiling and grimacing. Sometimes we can consciously control such instincts, and being able to do so is a sign of maturity and of intelligent self-awareness. We can mimic the facial expression of others because we have not only awareness of others but also the ability to mimic. Animals, too, have self-awareness and awareness of others to varying degrees, sometimes to degrees far exceeding our own (such as a cat or rat's sensitivity to familiar and strange odors). A number of animals have the capacity to mimic and often an understanding of the appropriateness of what they are doing, like the parrot who says "hello" only when people enter the room and "goodbye" only when they leave, and the dog who mimics a human grin only when he is greeting humans.*

People may demean animals and deny that they are intelligent and sentient (able to feel pain and to suffer) in order to safely distance themselves from animals. Then, through such rationalizations and defense mechanisms, they become immune from feeling responsible or guilty about how they (and society) abuse animals. Distance (and scientific objectivity) can sever empathy and responsible compassion. People don't like to suffer through empathy with the suffering of animals. Animal suffering (in needlessly repetitive research studies, in testing cosmetics and other nonessential consumer items, in the live trapping of wild animals for their fur, in the atrocious overcrowding of animals on intensive "factory" farms—the list is endless) is justified on utilitarian or economic grounds and rationalized on the basis that animals can't think, aren't intelligent or really aware, and don't really suffer, for when they are hurt, the pain response is just a reflex.

* This cannot be put down as simply mechanical conditioning. While conditioning is involved in such learning, there is also clear evidence of awareness in the appropriateness of the various responses in relation to the subtleties of different social contexts.

Concrete and irrefutable evidence from sound behavioral, psychological and physiological studies of animals gives us a very different picture. Many species of animals, and especially those wild and domesticated ones that we use to satisfy various needs, are intelligent, often rational, sensitive, ethical and altruistic. Being so, why should they not be incorporated into the sphere of moral concern from which to date they have been virtually excluded?

By establishing very close social bonds with animals, some scientists are at last beginning to prove the consciousness of animals more deeply and effectively than ever before. Dr. John Lilly has made some remarkable discoveries in his communication studies with dolphins, as have the Gardners with Washoe (who learned over 130 different signs and who invented some of her own) and other chimpanzees being taught to communicate using American Sign Language. Although it is beyond the scope and intention of this book to detail their research findings, it should be emphasized that it is remarkable what can be accomplished (with patience) once there is a close bond between experimenter and animal. With such a close bond, no experimenter can continue to perceive the animal as an unfeeling and irrational automaton. Through communicating with animals and exploring their consciousness, the investigator is *humanized* in so far as unfeeling, mechanistic attitudes toward nonhuman life are invalidated.

One anecdote from such research must be cited. A gorilla who had learned sign language was once asked which of the two keepers present he preferred. He signed back, "Bad question." One might infer from this that altruistic and egalitarian gorillas simply don't make such unfair social discriminations. While primates can be trained to make "if . . . then" logical inferences and to spontaneously make sign language sentences such as "Hurry, I want to go outside and play (or drive in the car or smell the flowers)," through this channel of communication researchers are also learning not only about their conceptual abilities but also about their self-awareness and ethical and moral sensitivities.

Certain views are being advanced by academicians that are relevant to how we, as a society, relate to and treat animals and nature. One such view is that humans are "superior" to other animals. Those who adhere to such a belief may then be incapable of ethically objective and responsible action toward animals because they perceive animals as being inferior. With this preconceived notion that humans are superior (and that some animal species are more or less superior than others) the ethically objective principle of giving animals equal and fair consideration—which is a basic premise of animal rights philosophy—becomes anathema.

In apparent support of this view, Harvard University sociobiologist and science popularizer Edward O. Wilson has constructed a list of the ten most intelligent animals. *The Book of Lists** in its section on the animal kingdom reproduced the following:

Dr. Edward O. Wilson's
10 Most Intelligent Animals

1. Chimpanzee (two species)
2. Gorilla
3. Orangutan
4. Baboon (seven species, including drill and mandrill)
5. Gibbon (seven species)
6. Monkey (many species, especially macaques, the patas, and the Celebes black ape)
7. Smaller toothed whales (several species, especially killer whale)
8. Dolphin (many of the approximately 80 species)
9. Elephant (two species)
10. Pig

Dr. Wilson adds: "I defined intelligence as the speed and extent of learning performance over a wide range of tasks. Insofar as possible, the rank ordering was based in part on actual experiments conducted on learning ability. In those cases where such studies have not been made, I relied on the 'encephalization index,' which measures the size of the brain relative to that of the body as a whole and has been shown to be roughly correlated with intelligence. Although I believe that my rank ordering is relatively sound, much more research is needed in this field of zoology, and changes in position can easily occur, especially near the bottom of the list of 10."

It should be pointed out, however, that "speed and extent of learning performance over a wide range of tasks," and the "encephalization index" are arbitrary, not absolute, indices of intelligence. Comparing the learning performance and brain size relative to that of the body of different species sets up absolute differences between

---

* New York: William Morrow, 1983.

Impeccable scientific research has now taught us that many animal species, cats included, are capable of a whole range of sensitivities and emotions, and are far removed from the brutish beings we once imagined them. Ongoing study may show levels and uses of animal intelligence that we still do not comprehend. *Photo courtesy Myanmar Birmans/David Hovell*

species. And when a hierarchy is drawn up, a further erroneous inference is made, namely that of superiority.

This "speciesist" thinking is a reflection of our own values, especially of our valuing intelligence as some special virtue. Such valuation can distort our perceptions of other animals and influence how we treat them and value them in and for themselves. A "dumb beast," low in the sapience or IQ hierarchy, would not be accorded the same respect as a more intelligent species (that is, one "more human" in some respects). Yet all animals should be respected equally, since they are all sentient, having the capacity to feel and to suffer.

While comparing different species is one avenue to understanding evolution, adaptation and the structure and function of living things, making comparisons on the basis of biased, human-centered values can have pernicious ramifications. Any hierarchy (of superior/to/inferior, greater/to/lesser) sets up a false view of reality, and when it is imposed upon the animal kingdom, it can break the circle of compassion within which all creatures should be regarded and treated with equal reverence and respect.

That Professor Wilson listed the chimpanzee first and not *Homo sapiens* is his first biological error. But it is, I believe, a politically coercive, if not unconscious, omission. It would seem by this omission that Wilson would have us believe that we are so superior to all other animals that we do not rank with them. Yet are we not, along with the gorilla and orangutan, less intelligently adapted to life in the water than a dolphin (which he ranks eighth)? And is not a tenth-ranking pig more intelligent at being a pig, and living in its own particular environmental niche, than a dolphin or a chimpanzee could ever be?

What is Professor Wilson doing? Other biologists before him placed Negro and European and Asiatic peasant races beneath their white, Anglo-Saxon Protestant ideal of superiority and human perfection. With such an arrogant attitude of patriarchal supremacy over others and the rest of creation, no ethical decision could be objective and unbiased. All moral choices would be made by some arbitrary consensus of reality and of our supreme place in nature that would be ultimately self-serving. Wilson's animal IQ hierarchy sets up the pyramid of power. Is it not as speciesist as his predecessor biopoliticians were racist?

Then there is philosopher Mortimer Adler, whose books, like biologist Wilson's, present the speciesist view of man's superiority over the animal kingdom as some irrefutable truth. His writings have wide public appeal and have been lauded by educators, though other philosophers generally ignore his pontifications.

In many of his writings he has argued that rationality is the highest virtue, and since man is the only truly rational being on Earth, he is superior over the rest of creation, and thus there is nothing morally wrong in his exploiting animals.

In his most recent popular book, *Ten Philosophical Mistakes,* he supports the views of Thomas Aquinas, who derived his philosophy from Aristotle and incorporated it into Christian theology, reasoning, for example, that only rational beings (that is, humans) have immortal souls. Animals are therefore inferior. So naturally, Adler is critical of Charles Darwin (who was cognizant of animals' emotions and was concerned about their widespread mistreatment, especially by vivisectors) for classifying man as an animal. Professor Adler contends that man alone can conceive of right and wrong. Thus man is a superior being, capable of moral responsibility and ethical conduct. However, we, unlike animals, have the power of free will to act immorally and unethically, so does this not make us "inferior" to other animals? That it is in our best interests to be morally right and ethically responsible is a sign of enlightened self-interest, not of superiority over the animal kingdom. It is worth noting that Adler stresses the difference in sapience of man over animals that makes him superior, rather than emphasizing (as did Darwin) the similarities in sentience—in emotional reactions and feelings—that make us feel kinship and compassion rather than superiority.

It should be emphasized that many opponents of animal rights philosophy have argued that only humans can have rights because only humans can act as moral agents. Nonrational animals, having no sense of right or wrong, cannot therefore have rights. But since babies and comatose patients, who are neither rational nor capable of being moral agents, have rights because they are recognized as "moral objects," it is illogical not to regard animals also as objects of moral concern with interests and therefore rights. Adler's thinking leads to the opposite conclusion, denying animals their rights.

In contrast to the unfeeling, "mechanomorphic" view of animals described earlier, the more sentimentally extreme *anthropomorphic* animals-are-little-people attitude may seem more logical and humane. But such an attitude taken too far can be just as bad (for the animals) as the mechanistic view. The anthropomorphic view can lead to excessive pampering and overindulgence that may be physically or psychologically bad for the pet. Very often a lonely person, or one who feels alienated from his or her own kind, will make a pet into a substitute companion person or substitute child. The owner may project all kinds of attributes upon the pet, verging on illusory and sometimes delusory notions. A wide range of emotions and thought processes that it does not have may be attributed to the pet. Even the animal's normal behavior (such as growling "Leave me alone") may be grossly misinterpreted (in this case as a "You don't love me anymore" response). Such excessive subjectivity in relation to the person's perception of an animal is clearly as unhealthy and as potentially detrimental to the well-being and "right" of the animal as is the overly objective and detached mechanistic attitude. Neither form of relationship appreciates the intrinsic worth of the animal.

We should also remember that many pets will learn to manipulate, indeed train, their owners to get what they want. Many cats and dogs have succeeded in training their owners to feed them only salmon or filet mignon, and are able to have most, if not all, their needs gratified—an owner may ignore guests to attend to the pet (who is jealous) or a husband and wife may sleep in separate rooms because the dog has decided not to let husband sleep with wife. Some pets will even pretend to be sick—for example, by feigning an injured paw—in order to get their owner's attention. We should, therefore, never underestimate the intelligence of our pets and, above all, never cease to regard them as being different from ourselves in terms of their individuality. We should respect and understand this "otherness" of the animals in our lives, and not, through anthropomorphic thought and action, try to make them seem more human than they naturally are. If we did, the significance of their "otherness" would be diluted and they might well be swallowed up in the sameness of anthropocentricity, a world where pets are mere artifacts, human creations that have no individuality but are merely extensions of ourselves. Let us preserve in our minds and lives their uniqueness and animalness, for on the other side of our "oneness" with nature and animals is the reality of their "otherness."

We can do more than simply feed, groom and play with our pets. We can begin by respecting them, not as our own creations or property, but in a manner more in line with the sentiment so eloquently expressed by Henry Beston:*

* *The Outermost House* (New York: Ballantine Books, 1971).

We need another and wiser and perhaps a more mystical concept of animals. Remote from universal nature, and living by complicated artifice, man in civilization surveys the creature through the glass of his knowledge and sees thereby a feather magnified and the whole image in distortion. We patronize them for their incompleteness, for their tragic fate of having taken form so far below ourselves. And therein we err, and greatly err. For the animal shall not be measured by man. In a world older and more complete than ours they move finished and complete, gifted with extensions of the senses we have lost or never attained, living by voices we shall never hear. They are not brethren, they are not underlings; they are other nations, caught with ourselves in the net of life and time, fellow prisoners of the splendour and travail of the earth.

Also, as we respect them, so we can explore their intelligence and consciousness, educate them and help them realize their potential under our custodianship. In the process, we will be educating ourselves, and by so doing, we will gain a deeper understanding of and reverence for our animal companions. Let us liberate them from the status of objects and property, from the tyranny of utilitarian exploitation and from our need to use them for selfish emotional gratification, all at their expense.

In animal liberation is also our own liberation, and as we begin to appreciate the intelligence and sensitivity of our companion animals, the day will come when they will be accorded the respect and rights that for too long they have been denied.

Even though the kitten shown in these photos is looking at a drawing and has no olfactory clues to indicate that the drawing is of a cat, the kitten's actions indicate self-recognition, and therefore demonstrable intelligence. *Photos courtesy M. W. Fox*

14

IQ Tests
(Games and Exercises)
for Cats

THE IQ TESTS and exercises in this book have been developed so that people cannot only evaluate their cat's IQ through the testing procedures per se, but also teach the animal something new. Consequently, its IQ will be improved and its conceptual and experiential worlds enriched.

There are also advantages to the owner and to the owner's relationship with the cat. A greater appreciation and understanding of the pet will come after observing and working with it. Cats need more than petting, feeding, cleaning and exercising. They need to be educated and entertained. Their lives can be enriched with IQ tests, and such learning experiences are rewarding games that can be enjoyed by the whole family.

It does concern me that many pets are often kept like household plants or toys that are fed, watered and occasionally played with. They can and should be given much more, and the more their lives are enriched with constructive IQ test games, the more our lives will be enriched with a deeper rapport and appreciation for our nonhuman companions.

This is, I believe, especially important for children. Being exposed to animals in childhood can give them a greater understanding of and reverence for life.

A child involved in observing, testing and teaching his or her pet some of the IQ test procedures described in this book will soon learn that cats are truly intelligent creatures.

Domestication has lowered the general intelligence and awareness of some of the animals we keep as pets. We should not demean them for this, since it was our doing in the first place. Some of the reasons behind the simplification and decline in behavior, intelligence and other traits as a consequence of domestication have been discussed earlier.

It is my hope that cat shows will place less emphasis on looks and stress more the importance of stability of temperament and the development of IQ, both of which can be tested objectively. Imagine how such tests could help improve the quality of many breeds. There *are* feline geniuses. People can do more than simply select and breed cats for good looks and placid (if not flaccid) temperaments.

All of the tests in this book are humane if done properly and require no surgical, drug or other "experimental" manipulation or intervention to alter the animal in some way. The scientifically minded student looking for a suitable (and humane) science fair project with an animal will find a variety of projects in this book.

Finally, the patience and understanding of the tester or educator are crucial. One who pushes the animal too hard, gives up easily or becomes frustrated and angry when the pet does not give the right response will have a significant negative effect on the animal's performance.

PRINCIPLES AND CONCEPTS

Before embarking upon a detailed description of IQ tests and IQ-enhancing lessons and games for cats, a foundation of basic principles and key concepts should first be laid out. With an understanding of some of the essential aspects of learning theory and practice, you can avoid mistakes and misinterpretations in the execution and evaluation of IQ tests.

First, let us consider *intelligence*. It is essentially the capacity to acquire and store information and to utilize such information (or

Domestication has lowered the intelligence and awareness of some pet animals, but if greater emphasis were devoted to developing these qualities through breeding and an enriched environment for them, people and pets would all benefit.   *Photo courtesy HSUS/McNees*

knowledge) at some later time. Acquisition is a function of *learning ability,* which in turn is influenced by several factors. Performance may be affected by *motivation*—in other words, by how keen the cat is to perform and learn. Hunger, fear, social praise and fatigue may influence motivation, either singly or together and either positively or negatively. A cat that is motivated to perform well in order to receive food or praise is likely to learn better than one that is not hungry or is afraid.

Motivation is affected not only by the cat's internal physiological and emotional states, but also by various *reinforcing* factors. Positive reinforcers include food reward and praise; negative reinforcement may entail punishment or lack of food reward.

For the best results, the physiological and emotional state of the animal should be as stable or as predictable as possible throughout all trials. If such is not the case, reinforcement may not be effective. For example, an animal that is not hungry or is too afraid may refuse food reinforcement. It will therefore have low motivation, its performance will be poor and any inferences as to its learning ability and IQ would be inaccurate.

In many intelligence tests and problem tasks, the animal may get rewarded in other ways. Certain actions—manipulation, exploration and even play—can be rewarding in themselves. This is called *self-reinforcement;* it is the enthusiasm of a subject who is eager to perform and to learn. Learning is coincidental to the task at hand, something that is often lacking in obedience training and in teaching (rather than educating) children in certain schools.

Various rewards may, therefore, operate in a learning task independent of the final consummatory reward and reinforcement of praise or food. Some "star" animals also seem to have a sense of satisfaction, if not accomplishment, in solving a particular problem. It should be remembered that *intermittent reinforcement* (not rewarding on a constant schedule) can motivate an animal more than constant reinforcement. Also, *positive reinforcement* is better for animals and children alike than punishment or negative reinforcement.

Many of the tests in this book entail hiding an object from the animal, who must then retrieve it and sometimes remember where it was hidden. To a very poorly developed nervous system, an object that has disappeared and can therefore no longer be seen or smelled does not exist. It has "gone." That the average cat will seek out a hidden object (and even cache one itself and later retrieve it) shows that it has what is called *object consistency*. The animal must, therefore, have the ability to maintain a mental image of the object that has disappeared (that is, it must have imagination). Psychologists refer to this as a *search image* (or *sollwert*). Without this ability, a bird would be less able to forage for insects under leaves and pebbles and a cat would have difficulty retrieving a toy it had cached under the sofa.

Retention is another important aspect of intelligence. Some animals have a poor ability to retain information. This is often, as with young animals, associated with a short *attention span,* the latter interfering with acquisition of information and with test performance.

Learning theorists have proposed that information that is learned must not only be retained (or stored, possibly as chemical RNA codes or *menmons*); it also has to be retrieved. It is difficult to evaluate which may be deficient—retention or retrieval—when a learned response is not correctly executed. Evidence from brain stimulation studies with implanted electrodes in humans points to defective retrieval or recall in lapses in memory. While a multitude of experiences are filed away in the memory banks of the brain, only a relatively limited number can be spontaneously recalled.

One must also consider the animal's individual (idiosyncratic), breed and species limitations and specialized attributes. A rat, for example, is particularly adept at learning a maze or complex runway, while a squirrel may do better on a place-learning test (since it is not adapted to living in tunnels or runways, but is adept at caching and relocating food stores in various places). A cat has a superb sense of smell and could outperform any human being in tests involving this sense, while a test involving manual dexterity could place a rat over a cat and a cat over a dog. Such *motor* and *sensory limitations* and special *attributes* should always be kept in mind when considering intelligence and learning ability.

Once an animal has learned a given task or solved a particular problem, the next question is how long can it retain (or remember) what it has learned.

How long an animal can remember (or retain) a given learned response is another important facet of intelligence. Basically there are two forms of memory, called *short-* and *long-term memory.* An animal may remember a complex sequence of actions or certain cues or signals for several weeks or months: this is long-term memory. An example of short-term memory is a cat's remembering where a toy or morsel of food has been hidden in a test in which the reward is concealed in one of two or more different places. Tests for long- and short-term memory ability will be described in the next chapter.

Sometimes recall can be facilitated by *association* or *déjà vu*. For example, a particular cue in an otherwise completely new context may trigger memories and a response elaborated earlier in association with a completely different set of cues. The similarity of certain past and present cues, by *association*, may not only facilitate recall but also enhance learning and the elaboration of some appropriate behavioral response. This is called *associative learning* and is probably the most widespread form of learning in animals other than conditioning (see later). For example, a simple connection between the presence of a particular person and receiving food reward is conditioning, but anticipating a similar response from another person is associative learning. Learning here involves certain similarities between particular cues or classes of stimuli (in this case, different people). The development of generalized phobias is a good example of associative learning, as when fear of thunderstorms generalizes into a fear of all sudden noises. The pattern here is one of responding to an increasing range of similar stimuli, which is called *stimulus generalization*.

Stimulus generalization in associative learning is also a normal aspect of learning, but the basic response pattern need not change. In other words, the response to a wide range of stimuli becomes habitual. While the response may be appropriate, *habit fixation* may limit the animal's abilities to elaborate new or more appropriate responses. Many animals show such habit fixation and a relatively limited repertoire of responses. Species limitations are particularly evident, as are individual differences as between one cat and another. One may continue to paw at a problem box (see later), while another may instead start to push with its nose or use its teeth. Perseverance and stereotypy of response in the first case contrast with greater flexibility in the latter. The difference may be attributed to a greater freedom from habit fixation, which allows the animal a new dimension for further learning and adaptive modification of its behavior in problem solving.

This freedom from always using a fixed or limited repertoire of responses is the freedom to learn and to elaborate new and more adaptable responses. A strong-willed cat, one that verges on the obsessive compulsive, may perform well in a variety of problem tests, but in others will, like a person of similar temperament, fail miserably. Being too "headstrong," it may make more mistakes, and worse, may

Demonstrating its intelligence and ability to learn, this cat knows from previous experience that the food reward is under the "O" card. *Photo courtesy M. W. Fox*

The cat being tested continues to seek out the bowl beneath the "O" card even though its position has been switched and the "X" card has been removed.

Photo courtesy M. W. Fox

not have the "cool" to pause and try out a new and potentially more appropriate response.

An animal that does have self-control and can inhibit learned responses to try out new and more appropriate ones is free to learn. Without adequate self-control or *internal* inhibition (as in a young animal, one that is overexcited or headstrong), reasoning and insight may be impaired.

Pavlov, in his conditioning studies with dogs, identified three basic temperament types. He designated a "weak and unstable nervous typology" to those dogs who had a timid temperament and were easily distracted and disturbed during tests. Dogs that performed well on those tests requiring action but did poorly on tests that required inaction (passivity or inhibition) were said to have a "strong but imbalanced nervous typology." These are the headstrong temperaments alluded to earlier. Pavlov's third type, his superdogs, did equally well on tests requiring action or inhibition: they had a "strong and balanced nervous typology." The ability to appropriately switch their behavior he termed *equilibration* or *dynamism*.

These landmark studies—which are also applicable to cats—reveal that temperament influences learning ability and IQ. To say that genes (inheritance) influence intelligence is too long a leap: the intervening factor is temperament or emotionality. While temperament traits are inherited, it has been shown earlier in this book how *early experiences* (the way in which it is raised) can help an animal develop a more stable and adaptable temperament. Good breeding alone will not produce intelligent animals.

Another important and subtle aspect of intelligence testing is determining the animal's ability to *unlearn* a given response pattern. This is an indicator of flexibility, not so much of insight but of the animal's capacity to control response perseverance or stereotype. For example, an animal that persists in looking for food in one place where food has been hidden earlier and cannot "switch off" its place preference and locate food hidden in another place is showing poor flexibility and high stereotypy or perseverance. A cat who continues to look for food under a card marked "X" when the food is now placed under card "O" is also showing maladaptive perseverance. The animal is unable to inhibit or "unlearn" its response and will always expect food under the "X" card. The ability to reverse behavior to different cues, as when one cue is first positive or rewarding and later negative or nonrewarding, is a good indicator of intelligence. Such an animal, in Pavlov's terminology, has dynamism or equilibration.

Four other types or patterns of learning remain to be described. One that is used in many of the IQ tests and for IQ development in this book is *learning sets*. With these, the animal essentially learns to learn. Through paced increments of experiences, the subject learns to master more and more difficult problems that are based upon the same learning principles. For example, learning to discriminate between different shapes or symbols can lead to discrimination between different groups of symbols or words until the animal is actually able to "read."

A common learning pattern in all but the least intelligent pets is *time-event* learning. The cat learns to anticipate certain occurrences during the day and, through conditioning and association, comes to know what its owner is going to do next. The animal will pattern much of its behavior after the owner's activities, such as waking up, anticipating being fed, played with, groomed and other routine activities. This is not difficult, since most animals (including us) are creatures of habit. Changes in such routines can upset some animals and may actually interfere with IQ tests. For example, it would be unwise to put an animal through its paces just after it has eaten or when it is anticipating a daily romp outdoors.

Many animals, especially cats, have the capacity for *observational learning*. After seeing their owner or another animal doing something, they may attempt to do the same. Also termed *mimic learning*, this capacity is highly evolved in the more intelligent animals such as humans and chimpanzees and is an important key to the acquisition of cultural knowledge, which is literally passed on from one generation to the next via observation and imitation.

Sometimes an animal may, in its behavior, display clearly rational and logical intent. *Insight* and *reasoning ability* are not purely human attributes. More highly evolved animals can also make insightful decisions and make "if . . . then" logical inferences.

For example, a cat may jump on a series of stools of increasing height to reach food hanging from a string that it cannot reach directly because the stool closest to the food is too high to get up on. By jumping on a lower stool first, it is showing insight and reasoning.

Observational learning may help an animal execute such an insightful solution, much like a young chimpanzee observing an older one using a trimmed twig to "fish" for termites. The observer, in copying, is making a logical inference: "If I do this, then I, too, should be rewarded."

Cats vary in temperament even more than they do in appearance. If you plan to try any of the tests in this chapter on your cat, be aware that your cat's individual personality will affect the scores it achieves. *Photo courtesy HSUS/Smith*

15

Basic Pet IQ and Temperament Tests

THERE ARE SEVERAL SIMPLE TESTS that you can put your cat through to evaluate its IQ. Some of the tests are games that your cat will enjoy and by conducting them, you may actually increase your pet's IQ by helping it learn to learn. But there are two great obstacles to be overcome before any such tests are attempted. First, the human problem. Only those people who are patient and who don't get frustrated easily should do these tests. If you are in too much of a rush or get frustrated when your cat doesn't respond or responds incorrectly, your cat will pick up your reactions and will become even more confused and frustrated itself.

The second obstacle is your cat's temperament. You should know its temperament before you try out the tests, because temperament has a profound influence upon performance. A cat that is afraid or disinterested won't score well. A poor performance does not therefore mean that you have a "dumb" animal, but rather, you or its temperament (or a combination of the two) are interfering with its performance. Knowing how to evaluate an animal's temperament can also help you pick out the best kitten in a litter, since research has shown that by six to ten weeks of age the basic temperament that an animal will have as an adult is already well formed.

Temperament testing can tell a great deal about a kitten's social responsiveness and whether that kitten is the right pet for a particular person. *Photo courtesy M. W. Fox*

TEMPERAMENT TESTS

These tests can be done on kittens from six to eight weeks of age onwards. In computing a kitten's rating, simply put down the appropriate score for each test and add them up. A very *high score* means a strong-willed and outgoing animal; a *low score* a shy, fearful animal. A *mid-to-high grade score* is probably the best, since such an animal would be outgoing but at the same time cautious and not foolhardy. These three basic "grades" correspond to Pavlov's three basic dog temperaments or "nervous typologies," the strong, weak and balanced types respectively.

Social Responsiveness

(a) When called, does the cat solicit your attention intensely (10) approach you slowly and quietly greet you (5) or shy away (2).

   Note down the number shown in parenthesis and continue with the other tests, similarly noting down the appropriate number for the response that the animal gives.

(b) If your kitten is with its littermates, does it push its way out first over its littermates to contact you (10), come up with one or two others to investigate you (5) or stay back and ignore you (2)?

(c) When petted or picked up, does the kitten get overexcited (10), remain quiet and relaxed (5) or "freeze" fearfully or tremble or try to escape (2)?

(d) When you quietly back away, does the kitten follow you immediately and solicit attention (10), pause and then follow you and solicit your attention less demonstratively (5) or go off and ignore you (2)?

(e) Call the cat to you, and when it is beside you, clap your hands loudly twice over its head. Then call it to come to you. Does it ignore the loud noise and continue to solicit your attention (10), cower or become passive and recover quickly (5), or freeze and refuse to approach even when you try to coax it (2)?

(f) How often does the cat respond to certain toys? Cats are often less playful as adults, so this test can give false measures with more inhibited or "cool" adults. Use a yard of string with a four-inch piece of towel or paper tied to the end. Drag it past the cat as though it were a mouse hopping by. Does the animal respond immediately (10) or paw tentatively or crouch and stalk first (5) or not respond except to look at the "prey" or simply ignore it (2)?

(g) How does the cat respond when it is in an unfamiliar place, say outside its pen or home, in a park or quiet yard? Does it explore actively but with some caution (10), or does it run around wildly exploring things and not calm down quickly (5) or does it freeze, cower or attempt to hide somewhere (2)?

Since my research has shown a relationship between heart rate and temperament in pups, you may wish to try to detect it in kittens, using a stethoscope. Pups with the highest resting heart rates in the litter tend to be the most assertive and outgoing, while those with the lowest rates are the most timid. The same may hold true for kittens, but this has not yet been researched. Simply hold the animal in your lap, and when it is quiet and not struggling, record its heartbeats for fifteen seconds; multiplying by four gives you the rate per minute. Outgoing pups have rates of 200–240 beats per minute, while others in the same litter having rates as low as 160–180 per minute will as adults most likely be shy and easily scared by sudden or unfamiliar stimuli. Those, then, will be the poor learners.

Since the foregoing temperament tests involve the presence of a person, which could bias the tests and lead to incorrect inferences, other tests without a person present are essential. So repeat test (g) with yourself concealed behind a blind or suitable screen. Next throw some large and unusual object near the animal, such as a ball of paper or cardboard box about the same size as the animal. This way you can test its reactions to *surprising* and *novel stimuli*. Other setups include a string pulley that suddenly pops up a cardboard shape in front of the animal, or approaching the animal and opening up a large umbrella. Score (10) for an unhesitating approach, (5) for cautious approach or withdrawal, then investigation and (2) for flight and no investigation.

Also observe how the kitten reacts, in your apparent absence, to others that it lives with or its littermates. Score (10) if it is always number one when playing with others or in getting food, (5) if it is intermediate and (2) if it is obviously the lowest one on the totem pole.

The above simple tests can help you select a kitten that will have a stable temperament, which in turn will be most amenable to training

As with puppies, there may be a correlation between kittens' resting heart rate and the nature of its temperament. There can be great variation here between very assertive and very timid individuals in the same litter.   *Photo courtesy HSUS*

and IQ evaluation. Don't forget, either, that training and testing are much easier if your pet is strongly attached to you. If it has not been properly socialized, you may not get much farther than simply evaluating its temperament.

HOW TO SCORE

Each of the following tests consists of five trials. For each correct trial, score 10, for each total failure, score 0 and for a slower, confused response that does eventually come out right, score 5. Add up the scores for each trial, multiply by four and you have an "IQ rating." Out of a possible 200, anything above 125 is superior, anything above 150 near genius. A score around 50 isn't so hot, and below that, you possibly goofed in conducting the tests or else your cat was either emotionally disturbed, simply not motivated or is brain damaged!

HIDE-AND-SEEK

This test is based upon the notion that when an object disappears from view, it can still be visualized or held in the animal's memory. This is termed object consistency. As with most of these tests, it is advantageous to have a helper who gently restrains the animal until you are ready for it to respond. You may wish to first "shape" your cat to retrieving or at least finding the toy by throwing it various places in the room. Once you have a consistent response pattern (say, five out of five), place the toy halfway under an old towel (crinkly newspaper may scare the animal) and repeat this once or twice. Praise the cat each time it responds correctly. Finally, place the toy completely under the towel and see what happens. A bright cat should be able to get the toy a few seconds after it is released by your helper. Don't do any more tests now until the following day, or your pet may soon tire of this game and give an erroneous low score.

Next day "show" the toy to your cat, who is being quietly restrained about eight to ten feet away. Place the toy under the towel and then back off to one side and signal your helper to release the cat. Repeat this test five times and calculate its IQ rating.

HIDDEN FOOD

Because not all cats will be motivated to find a hidden toy, the above test should be compared with your pet's ability to find hidden food. Show it some tasty morsel of food and place the food on a flat plate and hide it under the towel. Do this test about one hour before your cat's normal feeding time. If you test it at the usual feeding time, its habitual expectations of being fed (at the usual place and time) could interfere with this test, and, of course, after feeding the cat may not be the least bit interested in responding. The animal should make every effort to remove the towel and get at the food beneath it. Repeat this test five times and do not give it more than a quarter teaspoonful of food each time; otherwise it may get satiated too quickly. Comparing the results of this test with those of the first one will give you a good idea how food-motivated your cat is. A high score on this test can compensate for a low score on the first for an animal that isn't "turned on" by some toy object.

These two tests form the basis of several more advanced tests that may be conducted once you are able to get your cat to respond with a high degree of consistency. Because the animal may be able to see the shape of the toy under the towel, or smell it and the food, the next sequence of tests is designed to make things a little more difficult.

RECOGNIZING THE CHANGE

Wild animals that I have studied are extremely sensitive to changes in their familiar surroundings. Any novel object, a cardboard box, ball, balloon, crumpled pile of wrapping paper, and so forth, will trigger an immediate response: sudden attention, flight or cautious approach and investigation. Such animals are highly aware of their environment, and such awareness can be easily measured in our pets. With the cat in another room, place some completely new object in the middle of the living room floor, such as a large balloon or a small opened umbrella. Don't handle it much, or handle it with tongs, since your own scent could reduce its novelty. Then sit back and wait to see what your cat does when it comes in. For no response, score 0, for a brief look, score 5 and for a clear, orienting response, cautious or immediate approach and investigation, score 10.

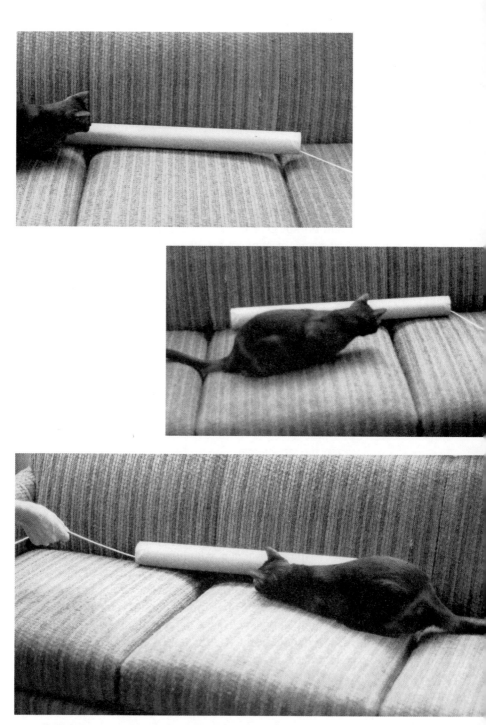

The tube test measures a cat's ability to grasp the spatial concept and how well it can anticipate and catch a toy being pulled through a long tube on a string.   *Photos courtesy M. W. Fox*

THE "TUBE" TEST

This is a special "cat-and-mouse" test that is very appropriate for cats and for most dogs as well. Get a cardboard tube about two and a half to three feet long with a three-inch diameter. Attach a small toy or morsel of meat to a string and pull it through the tube. Let your cat see it disappear into the hole at one end, and see you pull it out at the other end. Keep at least four feet away from this end of the tube. Your cat should quickly learn to come to the end nearest you to wait for the "mouse" to appear. Score 10 for a quick grasp of the spatial concept, 5 if you have to repeat it more than five times and 0 if your cat stays by the end of the tube where the "mouse" first disappeared. Having an assistant pull on the string at the point of entry into the tube can speed up this test. Then the meat or toy must be tied to the middle of a long length of string.

The cardboard cylinder can also be used for the more advanced "tool" using test in Chapter 16, in which the food or toy is pulled into the tube out of reach of the cat except for a wire loop or wooden toggle that the cat must pull on to pull the reward from inside the tube.

RIGHT AND LEFT TESTS

Use two cake (baking) tins and a cardboard box to cover the front of each tin. Your cat has access to either tin via the back of the box, which is open. Set the tins about three feet apart inside their boxes and show the animal in which box you are placing the toy. After the cat has made two correct responses, switch over and use the opposite box. It should not persist in going to the wrong box. Once you have accustomed your cat to this test with five or six trials, wait twenty-four hours and then test its IQ. Do five tests, placing the toy alternately in one box (A), then the other box (B), in the following pattern: A, B, A, B and finally B. This last repetition will show you whether the animal has developed a spontaneous A–B alternation, rather than solving the problem.

Repeat the above tests using a minute amount of the cat's favorite food. Smear each tin with a little of the food dissolved in water first to control odor cues. If the cat executes a wrong response on any of these trials, do not allow it to go near the other tin. Simply return it to your helper and remove the food or toy from the tin, hold it up for the cat

The cat that tests high in the tube test will be waiting at the other end to immediately pounce on its prey.   *Photos courtesy M. W. Fox*

to see and then place it in the next-in-sequence box, and when ready signal your pet to be released. For this test it is necessary to stand close to the boxes to prevent your cat from going immediately from one box to the next, unless you wish to make the boxes out of wood and have a guillotine door on the back that you can open or close at a distance with rope and pulley.

It is worth spending plenty of time on this test, because once an animal does understand what it is supposed to do, several more advanced tests can be built upon this one.

FRUSTRATION-DEXTERITY

Once your cat knows which box the toy or food is behind, the next test is to frustrate the animal by putting the food or toy on the ground with the tin inverted over it. Each tin should have one lip bent outward so that the animal can get a paw or nose under it. A dexterous cat that thinks first and doesn't give up easily or become overly frustrated and paw frantically at the tin should quickly solve this problem. Again, give five trials on one day and repeat the next day, giving five trials and scoring each successful response.

FRUSTRATION-DETOUR

For this test, you need a barricade of chicken wire or a couple of chairs with low cross bars on the legs that the cat can't get under. Place or throw the toy or food on one side of the barrier and then release the cat on the other side to go get it. Some headstrong cats will crash unthinkingly into the barrier. Those having more insight will immediately go around one side or the other. Repeat this test five times and record the number of correct responses. Repeat the next day with one side of the barrier blocked by a wall (put the chicken wire barrier or edge of one chair securely up against the wall). The cat should immediately go to the open side, or correct itself after one mistake. After the second trial, switch the block to the other side if possible. A corridor in the house is ideal for this reversal test.

SOCIABILITY AND DETOUR TEST

This same test can be conducted on kittens to evaluate their sociability and intelligence. Place the animal on one side of the barrier, step back and call it to come to you. Repeat the test five times, praising the animal each time it solves the detour and comes to you. An excessively aroused and distressed animal will at first keep pushing and pawing it the middle of the barrier to reach you. It should eventually solve the problem (keep both sides of the barrier open) and when placed back behind the barrier should come around with little further delay. Only start scoring once the kitten has come to you on its own accord on the first trial, and give five trials. Next day, repeat five times and then block one end, selecting that end first that the animal may favor coming around. If it is "stuck," show it the way out once only, and do five trials. The cat should not persist in trying to get through the blocked edge of the barrier. Then do another five trials with the other edge against the wall. On the third day, you can test the animal's ability to alternate its behavior by alternately blocking each end of the barrier (A-B-A-B-B).

Score 0 each time the cat cannot solve the detour problem, give 10 for an immediate correct response and 5 if it takes more than twenty seconds to come or if it tries to get through one blocked end first. Five trials will then give you a possible total of 50. Multiply your pet's mark by four, and you will have a rough score of the animal's IQ. A score anywhere over 150 means you have a very intelligent and sociable feline companion.

DEXTERITY-INSIGHT

Because cats are quite dexterous they are excellent subjects for this test. Have a cardboard box (with lid left open) about one foot high and two feet square, and let the cat see you drop some favorite toy or morsel of food inside. Let it jump into the box and retrieve it. Repeat this several times, then close the lid, with the food or toy in the box. What does the cat do now? It should try to open the lid, which should be easy enough for it to do. A slightly extended edge over the sides of the box will help it get its paw or head under. For an immediate correct response, score 10. If the cat fails, show it what to do and see if it gets the hang of things, since cats are generally good observational learners. Once the lid has been mastered, the next test entails cutting a two-inch-wide slit along the bottom of each side of the box, about eighteen inches long. Drop the toy (such as a catnip "mouse") or food into the box, close the lid and repeat; then put the toy or food in the box and cover the lid with a heavy board or tray. A clever cat will eventually try to get into the box via one or more of the slits, and will use a paw to reach in and get the food or toy. Score 10 for a quick, insightful response, 5 for a response after you have shown the cat the toy or food inside the box through one of the slits and it then responds, and 0 for no response.

We can do so much more with our companion cats than catering to their basic needs and training them to be obedient; we can, to a degree, actually educate them and develop their IQs. Scoring their responses and computing their IQs is part of the game for us. For them, it's the enjoyment of having human attention and engaging in a variety of problem-solving games. It's one way we can be closer to them and enrich their lives, and in so doing, we can help develop potential that is usually dormant because of the bland, unstimulating world we so often keep them in.

There is a host of behaviors most cats are capable of. Ordinarily, we never try to see what a cat can do. We may be surprised by what we can learn while tapping a cat's potential.

Photo courtesy M. W. Fox

16

Advanced IQ Tests and Exercises

THERE ARE several more advanced tests and lessons that can be given to your cat once the tests described in the previous chapter have been completed.

"TOOL" USING

It is possible to train cats to use a wire loop or piece of string with a stick "handle" as a "tool" in order to obtain food. What is needed is a gate (an expandable baby gate set up in an open doorway), the bottom bar of which is four inches or so off the ground. Place the animal on one side and a low pie tin containing a small quantity of the cat's favorite food on the other. Let the wire loop or string with a wooden handle lie on the animal's side of the gate, with the other end attached to the pie tin. Staying on the cat's side of the barrier, show it how to put the tin under the barrier, and do this for five days at the usual feeding time. On the sixth day, let the cat attempt to do it unaided. Score 10 for an immediate response, 5 if it tries but fails and 0 if the cat does nothing. Many cats will get the hang of this very quickly because they are naturally good learners through observation.

Before you do this test, you may want to check on your cat's *insight and spatial sense*. Have it chase a ball or other toy you throw for it; then throw it *under* the barrier. If your cat runs headlong into the barrier the first time, take note and score 5. If it runs into it the next time you throw the toy, score 0. If it immediately pulls up short and tries to reach the barrier on the first test, score 10.

SHORT-TERM MEMORY TEST

The following test is called the delayed response test and is excellent for measuring your cat's short-term memory. Two boxes and two cake tins are needed for this. Show the animal behind which box you have placed the toy or food in the cake tin and then have your helper cover the cat's eyes with one hand.

Its vision should be occluded for five seconds; then the cat can be released and allowed to choose the correct box. Do not allow the animal to correct any mistake *ever* in any of these tests. After it has made correct responses, the next day cover its eyes for ten seconds, and subsequently for even longer periods. A very intelligent cat can probably retain the image of the "correct" tin for up to five minutes. Score 10 for anything over two to three minutes, 5 for a delay of fifteen seconds to two minutes and 0 for any delay less than fifteen seconds.

VISUAL DISCRIMINATION

For this test, put a large "X" or other symbol on one box and a large "O" on the other, and train the cat to go to the "X" marked box in order to get its reward. Reverse the positions of the boxes so your cat doesn't get a right or left side preference and instead has to read the signs. Once you have trained it to do this, you may then switch it to respond to visual symbols, and you can build up its repertoire with a whole range of different cards.

Once you have your cat running to the correctly marked box, since it sees you putting food inside the box, it may still be responding primarily to you and not to the symbol. The first step, therefore, should be to give the animal five "shaping" trials and let it see you put the food behind, say, the box marked "X" each time. Alternate the positions of the two boxes from left to right: XO/OX/XO/XO/OX.

Next you must see if the cat has learned that "X" means food, and will only go to the "X" marked box. You can accomplish this by having your assistant block the cat's view as you place food behind box "X" and then switch the position of the boxes. Give five trials and allow the cat to correct its mistake if it goes to the wrong box. Then you are ready to give the test proper: five trials are allowed, and this time the animal is not allowed to correct itself. Keep score, and repeat the tests the next day.

A sophisticated improvement of this test can be constructed by providing a sliding guillotine door behind each box, operated by a pulley system. Keep the doors open during the first five "shaping" trials and then keep them closed during the next five. Gently pull the door open on the correct box when the cat goes to it, even after going to the wrong box first. Then when you run the actual test, keep both doors closed and open the correct one only when your cat goes straight to it. Don't forget to switch the position of the boxes between trials. This may sound complicated, but once you have learned how to do it, it will go very smoothly indeed.

You can make all kinds of variations on this test. Make it more difficult and advanced by using three cake tins with boxes, with two marked the same ("X" and "X") and the other marked "O." The animal must learn only to approach the odd ("O") box in order to get rewarded. Again, change the positions of the boxes to insure your cat isn't developing a position or place preference. Once your cat gets the hang of this, you can draw all kinds of different symbols for this oddity discrimination test. If you have carpentry skills, make some wooden cubes, pyramids and hexagons of different sizes, rest a set of three on plywood boards over three cake tins, and you are on your way to evaluating just how well your cat can detect the differences in shape, size and so on. The animal must learn to push the right object off the tin in order to get food.

In this kind of test, the cat is "learning to learn." The various trios of two similar and one odd shape serve as learning sets whereby the animal can learn to make finer and finer discriminations of size and shape differences.

You may wish to consider a slightly different setup for your cat using pie tins or plastic bowls not hidden behind a box but covered by a square of thin plywood or cardboard, on which you put the appropriate symbol.

Begin first with an "X" card over one tin and train your cat to push the card over to get a minute food treat under it in the tin. You must first let it eat out of the tin before you cover it. Then your cat must learn to push the card away with its nose or paw. Next, add another tin with a blank card, and let it work itself through the problem until it has the concept that food is only under the "X" card. After this has been learned, alternate the positions of the tins randomly from left to right.

Subsequently you may elaborate upon this test by putting a "O" on the blank card and again only putting food under "X." Once your cat is giving an 80 to 100 percent response rate, switch the value of the signs so that "O" means food and "X" means an empty dish. Work slowly on this reversal, since some animals may have setbacks due to "overlearning" or being pushed too hard. It is essential that you do not spend too much time on this kind of learning; about fifteen minutes a day is usually all that your cat will take before becoming bored or frustrated. Also always have "school" at the same time each day.

When your cat is responding smoothly to this card game, you will be ready for further variations. You may add a third tin and train your cat to choose the odd symbol ("O"/"X"/"O") or have five or six tins with blank cards to test for place learning, as described above using boxes.

The beauty of using cards is that your cat has to perform by nosing or pawing the card, while with the box, it simply has to go around behind it. With the cards, it is possible to train your cat to "read" and to "count." Begin with one card that has "CAT" written on it, while the tin next to it has a blank card. Say "cat" and point to the "CAT" card. Next put the word "FOOD" on the other card and train your cat to switch as it did earlier from "X" to "O." In this way you can build up quite a vocabulary and may eventually be able to set up three or four different cards and tell your cat to get "FOOD," "CHOW," "MEAT," "CAT," or "DOG." Similarly, you can teach it to count by putting one, two, three, four or more lines on each card. Remember, it is probably easier for your cat if you say "one two three" for the number three card with three lines on it, and so on.

PLACE LEARNING

This is another short-term memory test, which entails the cat's having to remember behind which one of several boxes (or under which of several pie tins) you have placed the reward. The test is best done outdoors or in a large room. Make a semicircle of five or six boxes each about two to three feet away from the next. Show the animal behind which box (or under which tin) you have put the reward, and have your helper release your cat as soon as you stand up (but don't look at the box where the reward is). Most cats will make one or two errors to begin with, and this test, like the foregoing visual discrimination tests, is not easy to score to get an IQ scale. The tests are simply good learning experiences for your cat and fun, too. Once the animal has mastered the six box positions, you can do further variations, such as adding more boxes or placing the boxes only a foot apart. Be sure that if you are using food, you have each tin smeared with a little watery extract of the food; otherwise the animal may get the right tin on the basis of odor detection rather than visual place memory.

PUZZLE BOXES

If you are an ingenious as well as a patient kind of person, you can try to build a series of puzzle boxes for your cat—and entertain visitors, too. First, build a box, say ten inches by fourteen inches by eight inches high, out of quarter- or half-inch wire netting so the animal can see into the box. Train it to go to the box and lift up a loosely hinged lid to get whatever suitable reward you have put inside. Once the cat has mastered the trick of opening the wire "cage," you can next hide the reward inside a small cardboard box or wrap it in paper. Another complication is to put a catch on the lid that the cat has to nose or paw up or to one side. Finally, you can put a wooden peg in the catch that the animal has to pull out before it can move the latch.

Since cats are dexterous, various devices can be put on the puzzle box to make it more and more difficult to open. One begins first with the box open with food inside. Then close the lid, which the animal must learn to paw or nose open. Then add a toggle pin to close a latch on the lid, so that the animal must pull out the pin before being able to lift the lid. Learning can be accelerated through observation—that is, you can show your pet what to do. So far we have a three-part sequence. It can be made harder by adding one or two more toggle pins, an extra rope around the lid that has to be pulled off, and so on.

All kinds of seemingly complex, but really rather simple sequences can be built up through this method. The extent is limited only by your interest and imagination and the understanding that IQ testing is not trick training.

If you are electronically minded, you may also wish to construct some operant conditioning systems for your cat. Many cats are good observers, learning by themselves how to switch lights on, open doors, refrigerators, and so forth. A low-hanging string and toggle attached to a secure light fixture with a pull switch will entertain many cats who obviously enjoy switching the light on and off. A similar setup can be made for an instant-playing radio. If you know how to set up circuits with photo cells and pressure transducers, you could arrange it so that when your cat walks, sits or lies in a certain part of the room, a light, heat lamp, radio or TV goes on or off. Animals learn with remarkable speed that if they sit in one place, a radio or heat lamp will come on. More complex systems can be built involving operant conditioning that help enrich the animal's environment and give it a sense of control and relief from boredom.

APPENDIX

Cat Queries and
Queer Cats

O VER THE PAST several years I have kept a file on some of the more usual and bizarre questions I am asked about people's cats. Chances are that one or more of the answers given in this chapter are pertinent to your pussycat because all cats are odd, unique individuals with intrinsic nature and worth warranting our deepest respect, especially when they are loved and understood for who they are.

BEHAVIORAL TRAITS

Here are some questions and answers that help clarify what cats do and why—more insights into "felinese."

Q. Is my cat a show-off or what? When we have company, she always parades, not walks into the room and then sits there in the middle of the carpet. Is she trying to tell me something?

A. You got the message exactly—your cat is parading, and I think she's showing off. Doesn't she always get lots of comments and attention from the other people when she pulls this stunt? Obviously,

No two cats are alike and all are unique personalities. "Catness" always encourages a flood of questions from those who share their lives with felines. Even steady, predictable cats can step out of character, prompting a plea for answers. *Photo courtesy Mary Pyles*

being stage center is a rewarding experience to her. My two cats do exactly the same thing. I get a kick out of observing them.

Q. Can you tell me why both of my cats like to roll around in the dirt?

A. Cats often like to dust-bathe outdoors, like birds that will roll and flutter and preen in a bath of water or dust. The dust or dirt is a kind of dry shampoo. This behavior may hark back to your cat's distant ancestry with the Libyan desert cat, which, I am sure, didn't have much chance to take a dip except in a "pool" of sand.

Q. My cat used to charge at mirrors, but now he stands there and hisses at himself. Same thing when he sees his reflection in a window. How can we get him to stop it?

A. What's wrong with a cat hissing at himself unless he attacks his mirror image? Many cats ignore mirror images. Others make a game of rushing at the mirror and scaring themselves into a panic after giving a defiant back-arch and hiss. This implies that cats *do* have imagination, some perhaps more than others. I have seen kittens explore behind a mirror, obviously trying to locate the other cat. One who can't may well be just a klutz. Let your cat enjoy himself.

Q. I have a part-Persian cat and find him a joy. I have a question. When a cat flips and waves his tail, is it a sign of friendliness (as in the case of a dog), or annoyance or impatience?

A. In my book *Understanding Your Cat,* I give detailed information about feline body language. Waving the tail slowly from side to side, and especially rubbing it along something, can be a sign of relaxed friendliness. When the cat is impatient, annoyed, or getting ready to pounce on prey, the tail may be lashed briskly or waved vigorously from side to side. When the tip is wagged to and fro, it's a social signal that says, "Yes, I'm listening to you or know you're there, even though I'm not looking at you."

The various tail positions that cats display, therefore, have very many different meanings. Sometimes the same signals, occurring in different contexts, can mean very different things; so there is some ambiguity in the signals depending on the existing circumstances.

Q. We feed our cat indoors, and she sometimes tries to bury her food under her mat! Why? Also, when I play with her, she may scratch

A "loving session" is a form of social grooming for a cat, and prompts oneness with a special human friend and a general feeling of calm and well-being.   *Photo courtesy HSUS*

me. If I hit her gently and say, "No!" she bites my hand and settles down. The next minute she starts licking my hand! Again, why?

A. Yes, some cats will use whatever they can to "bury" their food in—a mat or a towel. They may even upturn the bowl to cover what they don't feel like eating. Your cat's response to discipline is quite normal. The bite means, "Okay, I can hurt you, but I don't mean to. See?" And the lick means, "Let's be friends and groom each other." Once you understand "felinese," life will be much pleasanter for you and your feline friend.

Q. Why does Baby, my cat, want to lick my hands, my arms, and sometimes my face, just like a dog?

A. Your cat is grooming you affectionately, just as she would a kitten or a companion cat that she likes. Taking care of each other is part of loving. Aren't you doing the same—reciprocating loving attention when you stroke or pet your cat? Such gentle grooming in cats causes total body relaxation in them. The heart rate slows down, and a general sense of well-being is created. This social grooming in cats is different from the face kisses dogs give, which are just social greetings. But dogs, too, will lick and nibble each other when they lie down together. The more they (and we) touch, the more they enjoy and express their affection for each other, and for us. Touching is loving!

Q. When we contacted our local school of veterinary medicine for an answer to this question, we were told by the professor to write to you. Our question is: Do cats and dogs have dreams?

A. Yes, indeed! Recordings of brain wave activity in cats and dogs that are dreaming are identical to those of human beings in a similar state. From this we can conclude that cats and dogs do have the same physiological state seen in man during the dreaming phase of sleep. In addition, body and limb movements—twitching of eyes, ears and whiskers, and occasional vocal sounds—clearly indicate this stage of sleep in man and all animals.

Q. Our three-year-old cat has a ragged cloth toy that he carries around the house, making buzzing and purring noises deep in his throat, followed by plaintive cries. Other habits connected with this toy are bringing it to me for throwing, so he can fetch it, and keeping it beside him when he rests, sleeps and eats. I find it touching. How do you as an animal psychologist view these actions?

A. Your cat's behavior is interesting, but not unique. The vocal sounds you describe may be the particular call of a cat when it catches the sight of prey, such as birds or mice, or when it's calling the kittens to "come and get it" after the prey has been caught. Cats will also retrieve prey and bring it home to feed their offspring, and this behavior is sometimes redirected to the owner in a form analogous to giving the owner a token offering!

You should remember, too, that cats are social, and that the fetish object to which your cat is so attached is not only a substitute prey but also a substitute play companion as well. It could also be a substitute kitten, and you are witnessing your cat's paternal and protective side. The main thing is that the toy keeps your pet happy, so let him enjoy this little vice.

Q. I have a problem with my calico cat. Since early spring she has eaten a few of the small animals she catches—for instance, birds, mice and even a baby rabbit. What is the reason for her doing this? She's about six years old, eats mostly cat food, with some boiled liver approximately once a week.

A. Most cats have a strong instinct to go out hunting. In domesticating the cat, selective breeding has done very little to change this natural instinct, and until recently cats were generally kept for the prime purpose of holding down vermin around the homestead or farm.

Dogs have been domesticated for a much longer period and have been selectively bred for a greater variety of tasks, to the point that some dogs show no interest in hunting at all. Most cats, however, still show this natural instinct, especially if they had some experience with live prey in kittenhood, but even this is not always essential to make them into effective killers. This is the reason I advise cat owners to keep their cats indoors. And applying a bell to a cat's collar doesn't always help. A cat can learn to walk so that the bell emits no warning tinkle whatever!

Q. Our two-and-a-half-year-old cat, Tina, responds to flashlights and reflection. We first noticed this when we were using a flashlight to clean out a closet. She chased the light, and the next night started hunting for it. We now use it for her entertainment. Should we limit these play sessions to very short periods because of her extremely elevated activity?

A. Thank you for your letter. What a good idea for night games

Camouflage works as a hunting tool for domestic cats as well as it does for their wild relatives. The gray and white coloring of this cat helps it blend into the shadows as it patiently awaits prey movement in the grass.

Photo courtesy Mary Pyles

The cat is a wonderfully evolved hunter, and everything in its nature is geared to this purpose. Even after centuries of domestication, the cat you share your life with, like this urban panther, is still instinctively close to its feral ancestors.

Photo courtesy M. W. Fox

181

for pussycat owners! And how about a mouse or bat silhouette over the front of the flashlight to focus on the wall? You're right to be concerned that she doesn't get overstimulated, but only because she might knock down ornaments and such as she goes up the wall, or gets so turned on that she won't let you go to sleep! Otherwise, let her enjoy!

I believe that some alert and intelligent cats get bored being indoors all the time and need some creative games. They even invent their own games; chasing light reflections from a hand mirror reflecting sunlight on a dark wall, for instance, is one way to compensate for a bland domesticated existence.

Q. I have twin cats, two years old, who haven't been out of the house since they were born. Each morning when they wake (at about 5:30 A.M.!) they immediately start scratching on all the mirrors and pictures hanging on the walls. They chase each other around, up and down furniture and beds for about five minutes. Sometimes they go through this whole bit again late in the day. Can you tell me why?

A. What you're describing is "the early-morning crazies." My own cats do this around 5:30 to 6:00 A.M. It's really very annoying, but you just have to live with it. You see, cats are crepuscular, that is, they are usually active in the early morning and evening. What you're seeing is a natural rhythm of behavior; in the wild, your cats would be active and go out hunting at these hours.

Even the fact of being raised indoors, and years and years of domestication do not seem to have eliminated this natural rhythmic behavior. Birds raised in zoos, for instance, become more active at that time of year when they would normally migrate. This whole area of rhythmic activity in animals and people is indeed fascinating, and maybe you can better stand being wakened at dawn by your "crazy cats" now that you better understand why they do it!

Q. My nine-year-old cat sleeps on his back a lot. I have heard that it could be due to worms, but Tommy has done this since he was a kitten. I've had him to the vet and I got pills for him, but he still sleeps on his back. Would appreciate any advice you can offer.

A. What pills did your vet give you? Do let me know, since I can't believe that a D.V.M. would offer treatment for your cat's non-problem, since no treatment is needed! A relaxed, warm and happy cat will sleep on its back. Worms have nothing to do with how cats sleep.

I've heard many myths about what worms can do to pets, but this

Understanding the natural habits of a cat make it easier to live with. Your cat may be dormant all day but experience frenetic spurts of activity during morning and evening. This is the natural rhythm of behavior in action—undiminished even after centuries of domestication.

Photo courtesy Mary Pyles

is a new one! Are there any other good ones that readers can tell me about? And how about where worms come from? One reader told me they come from candy.

Q. My cat, Tinker, kisses me whenever I ask him for a kiss. Is that normal? Also, whenever I hold him, he nibbles my hand. Is he trying to tell me that if he wants to bite me, he can, so I should watch my step? Is it normal for an altered cat to cut through the house at ninety miles an hour, jumping over everything in his way, and then come to a complete stop and go to sleep five minutes later? Also, Tinker goes around talking to himself while cutting through the house and climbing the walls. Are all these things normal?

A. You have a "crazy" cat who is all cat, full of fun, obviously well-adjusted and bursting with *joie de vivre*. Happy cats will often do seemingly crazy things, like racing through the house, hallucinating, etc. Cats will kiss, or more correctly, touch noises, almost any time a person bends down to them. It is perfectly natural social behavior, much the same as your cat's biting you is a sign of play. Try biting your cat back in return—on the cheek or paw. Crazy cats like crazy people, and in my bag, "crazy" means A-OK.

Q. Why don't cats kiss like dogs when they greet you? My cats lick me when they're being stroked sometimes, but never "kissy face" like a dog.

A. The usual greeting for a cat is simply a nose touch. They lick each other only for social grooming. The dog, however, not only licks to groom a companion. Its face-oriented licking (or what you call kissing) is similar to what it did as an infant. This kind of licking behavior solicits the attention of the mother and will often make her regurgitate food. The cat's equivalent of a kiss is to touch nose to nose.

So the developmental history and evolution of face-licking in the dog, perhaps similar to mouth-to-mouth kissing in man, is actually derived from an early solicitation of food from the mother. Cats, however, do not regurgitate for their offspring, and this is probably why face-oriented licking (kissing) has not evolved in them. Guess you'll just have to be happy with kisses from only one of your pets!

Q. My cat will suddenly bite and scratch and then run off sometimes when I'm petting him. Is he schizophrenic or what? It seems as though I can't trust him anymore.

A. Some cats will suddenly attack while being petted, or withdraw abruptly, clearly showing that they don't wish to be touched any more. These sudden changes in temperament are often misinterpreted as being, variously, a mean streak, wildness or schizophrenia, aloofness or unfriendliness. In the first instance, the cat, as it relaxes, (especially when it is on its back while its belly is being stroked), may suddenly feel vulnerable and act defensively by scratching or biting. The seemingly "aloof" cat withdraws because stroking can be overstimulating and actually uncomfortable when the animal is tense or has been fully charged with a few strokes and has had enough. So don't overstimulate your cat, and when it is in a potentially vulnerable position, avoid contact!

SOCIAL BEHAVIOR

The social behavior of the cat is a cause of much misunderstanding and problems. Here is a sampler.

Q. Both my friend and I are veterinary technicians, and both of us have a peculiar problem with our cats. Andrea has eight cats and I have ten. We both love and understand the feline species, and they have everything a cat could want. Four months ago Andrea's favorite cat, a Siamese spayed female, took off down the block to live with a neighbor. And one month ago, to my dismay, one of my spayed females took off two miles to a suburban area and banded with a bunch of strays. Now she is out in the cold all the time, is fed by a nice lady, but gets no affection and is not allowed in the house. I took her back home, but the next day when I let her out, she went back to the same place! (She obviously knows where she is going.) Do you think too many cats can cause a personality problem? We always felt that we had insight into the feline mind, but these rebuffs have set us back.

A. Don't take it as a personal "rebuff" when one of your cats takes off and finds as crash pad somewhere else. Many cats like to have a relatively solitary life-style, and your homes may be just too tense and crowded for the more introverted feline personalities. Sometimes the runaway is an outcast that is bullied or ignored by the others or has no buddy cat to groom and sleep with. Other times it's an independent and dominant cat who doesn't like the status quo of the group. You should keep a close eye on the social relationships in your "pride" of cats, and if a newcomer doesn't integrate within a week or so, do your best to find a home for it. Some will escape and become feral, living more less independent of people. (The feral cat population in many areas is a serious problem.)

Q. From the time I brought our male cat home as a kitten, he has nosed our then year-old female cat out of her feeding bowl. Result: I have had to feed them in separate rooms, with the doors locked, so that the female could get her proper nourishment. Is this a matter of regarding herself as a mother and thus indulging her ''offspring''? Or is this an expression of male dominance and chauvinism? I notice that in their sham battles she lies on her back while they each make the most frightening threats. But when the cussing is over, it is he who chases her through the house. Do you have a clue to their peculiar behavior?

A. There's nothing at all peculiar about your cats except that what they do is peculiar to most cats. My own pair behave much as yours do. The female will back off if the male looks at her when she is eating, so they have separate food bowls and eat in separate rooms when there's a special treat such as shrimp or scallops leftovers. Your male grew up to be dominant over the older female, but dominance in cats isn't always absolute. Your female may dominate the male in other situations, such as a favorite resting place. What you are seeing are facets of the feline's complex and fascinating social behavior.

Q. Regarding your reader's letter about the cat who ''bites and scratches and doesn't want to be held or petted,'' may I suggest one possible answer? This cat may be one of the many who simply doesn't like to be picked up and cuddled. They enjoy being petted on the floor, on a sofa, or anyplace where they can keep their feet on the ground, and they may thoroughly enjoy romping and playing with a person. But lots of cats just don't go for the grab-and-cuddle bit. Personally, whenever I see someone trying to cuddle an unwilling cat, I am reminded of how children cringe when an overdemonstrative relative tries to smother them in hugs and kisses. Remember how awful that used to be?

A. Yes! And thanks for your useful and sensitive observations, expressing accurately what I believe to be a correct interpretation of those felines who freak when picked up. Early training in kittenhood, getting a cat used to being picked up and enjoying contact can help significantly in this regard. But most important are people's expecta-

If your cat protests when you try to hold and pet it, yours may be one of those cats that objects to such attention. There are other ways to exchange affectionate gestures, and it is worthwhile experimenting to find the ones your cat likes best.   *Photo courtesy Mary Pyles*

tions. I wish people would learn to respect the fact that some cats don't like to be picked up and hugged, and that they really have a right to protest.

Q. What can I do about my four-year-old female Siamese cat who refuses to accept a new kitten in the family? I've had the new five-month-old male for two months, but the Siamese doesn't seem to adjust. I separate them as much as possible, because the Siamese hisses and growls violently when they're together. I've also noticed that the Siamese is no longer affectionate and will not allow anyone to hold or touch her. Hope you can help!

A. Sibling rivalry combined with the relatively antisocial behavior of cats gives you only one option: find a home for the newcomer and try to make it up to your other cat. Siamese are sometimes very possessive of their owners, and any threat to that close bond means trouble and emotional distress for the overdependent puss. Anyone wanting to introduce a new cat when they already have one should do it on a trial basis. If, after three or four weeks, they haven't made amicable adjustments, then move the newcomer out, or the resident cat might move on if given half a chance.

Q. I wonder if you ever ran into a problem like this. I have two altered male Persian cats, three and six years old. The younger one is driving me up a wall. Every once in a while he acts like a female in heat, rolling on the floor and crying like a cat on a hot tin roof. Whereupon the older cat jumps on him and they go through the motions of breeding. I could never stand a female in heat. What is going on here?

A. Are you 100 percent sure that the rolling, courting Persian male is a male? I once saw a cat come in to be castrated, and it was a female! She was returned to the owner, who wasn't given the full story. "Jimmy" grew up and surprised "his" owner with a litter of kittens. So please have your vet check your cat. There are hormonal disturbances in a male cat that could cause such symptoms, a hormone imbalance or a tumor producing female sex hormones. If there is nothing wrong and he is a "him," let me know. I have seen male dogs who have developed rather special homosexual bonds, but never once have I heard of cats having a homosexual relationship in which one male is obviously playing the "queen." This could, however, be the case with your young cat, since his submissive reactions to the dom-

Introducing a new kitten to a household with an older cat should be done on a trial basis. There is no guarantee that the older animal will accept the newcomer. If they can achieve the harmony of this pair, all's well. Otherwise, be prepared to place the younger cat.

Photo courtesy HSUS/Rodgers

inant cat may trigger the latter to mount and bite his scruff, partially as a display of dominance and partially as a sexual response to the younger one's display.

Q. When my cat was younger, he used to get into fights, but they were never bloody! Now that he's seven years old, Pussy comes limping home with his front leg scraped and the fur torn off. We then try to keep him in the house, but he meows so that we have to let him out again. What can we do? His homecomings are getting more and more horrible.

A. It is sad when an older tomcat who is no longer king of his neighborhood comes home month after month all chewed up. Cat bites often result in serious abscesses and slow-healing wounds. Your cat, because of his sex, is driven to go out and fight and be attacked by rival males. It would be a kindness to have him castrated. This works out well with many tomcats, and they become contented homebodies, no longer driven compulsively to go outdoors to roam and fight to prove their "cathood"!

Q. We have two virile Siamese males and one fixed female stray. The two males got along well for eight years. When we brought the female in, things were peaceful for a full year (except that the males began spraying more and more). But then the two males started fighting, and if left together, I think they would kill each other. Is there anything that can be done to restore harmony?

A. I hope your letter will serve to warn cat owners that two cats are company for each other but three often mean trouble. You may be able to smooth things out by having your male cats neutered, but considering their age, that may be risky. Another solution is to find another home for one of the males or to get rid of the female who is the cause of all the fighting. Even so, the tomcats may still dislike each other, even when she's gone. I'm sorry that I can't offer you a better solution. Your problem is one of the behavioral problems where prevention is easier and better than the cure.

Q. My husband and I don't have company that often, but when someone does come over, my cat will hiss and growl. She will not let anyone touch or come near her. Is she afraid of them?

A. Some cats simply don't like to be petted by just anyone. They refuse to be sociable hostesses or cuddly kittens with people they don't

The mating act in cats is characterized by considerable sound and fury. Copulation actually triggers ovulation, and the male cat restrains his vocal consort by the scruff of her neck as he ejaculates.

Photo courtesy M. W. Fox

know. You should respect your cat's choice. In fact, you might want to take a closer look at some of your friends. The dislikes that some animals have toward people can sometimes be quite perceptive.

Joking apart, I deplore people who insist upon dragging their dog or cat over to the visitors to be friends. Some visitors couldn't care less anyway! Let the pet come over to be introduced when it feels like it. There's no need to abuse them the way some people abuse their children.

SEXUAL AND OTHER BEHAVIORS

The sexual and infantile social behaviors of cats evoke many queries that reveal how much cat owners can learn about cat behavior and psychology when they know what to ask. The sexuality of the cat and its proclivity for nursing seem to have qualities of perverse behavior, but this is in fact a serious misinterpretation.

Q. When my blue point Siamese gets really affectionate, he will often give me a bite—a very gentle bite. What is he saying to me?

A. He's saying "I love you, my friend." During courtship, a cat will seize the skin at the back of the female's neck and gently bite and hold it. I had a Siamese cat that used to give me a love-bite on the nose, and I would respond by gently biting a front paw. Interestingly, the harder I bit, the harder he would bite. How's that for subtle communication?

Q. Our cat adopted us in November; she was five months old. She has always been dainty and fastidious. In June, looking me straight in the eye, she backed up to the brick wall and sprayed! Twice since, she's jumped up on the refrigerator and sprayed the cupboard doors. What's with her? Is it because she has not been spayed?

A. Surprise—the female of the species can sometimes spray with just as deadly aim as the male! This kind of marking behavior could be related to your cat's having reached full psychological maturity and her need to establish her own personal territory and sense of security. Possibly, there's a strange animal in the neighborhood threatening her security!

Yes, you guessed it (or you've been reading my column closely). Spaying might help get rid of this nuisance behavior. You have a

The neck-hold used during copulation is also a highly effective means of inducing immediate submission. Mother cats use it when moving very young kittens from one place to another, and people sometimes used it to handle cats when there is a chance of the handler being scratched or bitten. *Photos courtesy M. W. Fox*

fifty-fifty chance of its working, so why not try it, particularly since it will also make life easier for her in the long run.

Q. We have a beautiful cat. He is eight months old. At times he looks at me with a sort of baleful expression. When I see that, I get prepared. If I'm walking, he jumps up at me; if I'm sitting, he bites me. What causes this, and what can I do about it?

A. Without further details about your cat's behavior, I can only offer two interpretations. One is that the baleful look and behavior as you describe are, in fact, sexual. The cat is regarding you as a sex partner. But, of course, he is somewhat confused by your size and the fact that you have two legs to walk on instead of four. The bite, therefore, if it's not painful, but playful, is analogous to a love-bite. The other possibility is that your cat is regarding you as a substitute prey object—something to play with, chase, ambush and so on.

A solution to the first possibility is to provide your pet with a companion cat—or, if all else fails, have your cat neutered. (In the interim, discipline might be quite effective.) If his behavior is due to the second possibility, you should give your cat lots of objects such as balls and fluffy toys for it to catch, pounce upon and play with. You might explore these various approaches before you decide on neutering. Good luck!

Q. I was going to get my one-year-old cat spayed, but friends have told me I should let her have one litter of kittens before doing so. Is this true?

A. Some people claim that allowing a cat or dog to have one litter of babies results in a better temperament. There is no evidence for this, however. I advise people with pets to have them neutered before or after the first heat, depending on the opinion of the veterinarian who's going to perform the surgery. I also contend that no cat or dog should be allowed to breed unless it is a very superb specimen of a particular breed or variety; there simply are too many cats and dogs around today that have to be needlessly destroyed!

Q. I have seen cats mating, and it's all so noisy, so fast and violent, like rape. Is this normal—what's going on?

A. Cat lovemaking, to a human outsider, does seem something out of the Marquis de Sade's imagination. The screaming or caterwauling comes from males vying with each other, possibly establishing which is "top cat" by the intensity and timing of their screams. Such ritual vocal fighting saves combatants from injury that might result if they engaged in physical combat—akin to the shouting matches of highest-ranking teachers of Oriental martial arts. The visible effect on the contestant of a perfectly timed and attuned shout— "Haaii!"—is sufficient to decide the match. So it is with tomcats, I believe. Such contests may do more than entertain the female. She may also be brought to a high level of sexual receptivity. Winning for the male does not guarantee him the female. She may reject him. She signals her acceptance with a specific call in response to his mating call to her. She does not ovulate until the male stimulates her, this mechanism (of reflex ovulation) greatly increasing the queen's chances of becoming pregnant. The male stimulates her intensely, because his penis is covered in small, backward pointing barbs. Hence the "act" is short, if not sweet.

Q. My black-and-white mother cat had kittens (all tigers), and eleven days later her daughter (a tiger) had kittens (all black-and-white). The mother cared for her daughter's kittens, just as though she were having kittens again. Now they all are one big happy family. I doubt that the kittens will know which cat is their mother or either mother know which kittens are hers, and little do they seem to care. Does this happen often?

A. Yes, this is not unusual when mother and daughter cats both have kittens around the same time. It can get even more complicated sometimes when there is a third adult cat around. I have heard of one or both nursing mothers allowing one or more of their adult offspring to nurse off them!

In the lion pride, two or three lionesses will share their litters and nursing chores. Such communal nurseries are a fascinating aspect of animal behavior. No less remarkable are the adult "baby-sitters" who may be left with the offspring in a safe place while the rest of the family goes off for a while. This occurs in the case of wolves, cattle and many other animals, proving the old truism that in many ways they are no less like humans than we are like animals! Domestic cats, like wild bobcats, will often nurse their kittens until they are eight to nine months of age.

Q. We have a beautiful kitten named Tom. The problem is that he nurses on everything! At first he would try nursing on the clothes we were wearing, and we could only stop him by wrapping him up in a blanket. Now he will suck on the blanket (or almost anything). We have tried pushing him away and giving him milk, but nothing works. What can we do?

A. Let him be. Being weaned too early probably aggravated the problem. He may grow out of it. Give him his very own blanket to do with as he pleases. Lock up all other items he may want to suck on, and when he sucks on you, discipline him. Otherwise, he may become neurotically overattached. Weaning your cat from this behavior problem now while he is young will save you continued trouble, and it will be much easier than trying to stop him when he's older and the habit is really ingrained. Be firm, and loving too!

Q. My daughter's cat, Ice Cream, had three kittens about five months ago. We gave one away. The two kittens we have seem to be healthy, but there's one thing about them that is quite odd. For twelve weeks Ice Cream nursed the kittens, but since she stopped, they have been nursing from our female dog, Daisy! Daisy is about four years old and has never had puppies. Daisy loves the kittens and doesn't seem to mind their nursing. Have you ever heard of anything like this before?

A. A mother cat will usually give disciplinary and inhibitory signals to the kittens when she is ready to wean them. Your dog, however, may not have the right signals to switch the kittens off, or she may be an overindulgent foster mother. If she is producing milk, which could be related to a false pregnancy or to the kittens' stimulation, it would be advisable to have her spayed. This could prevent certain complications in her uterus or mammary glands that might otherwise develop. Although this is not a wholly abnormal behavior, I would protect the bitch from her own strong maternal instincts and keep her apart from the persistent kittens. Or I'd have the babies adopted if they persist in nursing much longer! Perhaps your gentle dog just can't say no!

Q. I have a mother cat and two of her kittens (now one year old). What do you think of my never letting these cats go outside? They have all been neutered, but the male cat looks out the window at night and yowls. Am I being cruel to these creatures by depriving them of their freedom to climb, roam and investigate?

A. No, you are not being cruel keeping your cats indoors. You are being responsible and kind. If they don't get out, they won't get

When it is time to wean her kittens, a mother can serve unmistakable notice that the time for this transition is at hand. It is unusual to find queens as indulgent as this one who will let older kittens nurse with impunity. *Photo courtesy HSUS/Kostecki*

into fights, be injured by larger cats, run over or poisoned. Nor will they kill birds and other wildlife, or mess on people's porches and flower beds. It's very easy to take your cat for a walk in a backyard or garden—they enjoy it—but use a regular collar and a long string, and let your cat lead you around. The truth is that cats do well indoors, and there's no reason ever to let them out. And if they have always lived indoors, they probably won't want to go out anyway!

However, cats need fresh grass for their digestion, so you may want to grow a boxful in your window. If you choose to pluck it from outdoors, just make sure you wash it well. Also, be prepared, just in case one of your cats gets out accidentally, and have collars and identity tags put on them. Use collars that have a built-in safety elastic expander. In that way, if your cat gets hung up on something outdoors, it will be able to wriggle free.

EMOTIONAL PROBLEMS

Cats suffer from a variety of emotional problems that are expressed in various ways. Being able to recognize a change in behavior as a possible sign of emotional distress (rather than misinterpreting it as sheer cussedness or stupidity) can do much to save cats from avoidable suffering.

Q. My cat spits and hides when company comes. She's a lovable cat with me, but she's impossible with visitors. I feel embarrassed and kind of rejected, too. People think I'm weird living with such a spook. Please advise.

A. A veterinarian friend of mine, Dr. Mark Applebaum, aware that cats do develop xenophobia—fear of strangers—insisted that all visitors play with his two kittens as soon as they sat down. The result is two well-socialized, emotionally adjusted adult cats who make everyone feel welcome. Sorry, it's too late for you to do this with your cat, but next time, start out right with a new kitten and follow my friend's routine.

Q. Knowing of your reputation as an animal behaviorist, I am seeking your advice. One of my clients has a neutered eight-year-old Siamese cat named Seuss. Last summer, business took the woman out of town frequently. Seuss was kept in the apartment and fed by a neighbor. Then he started acting strange—pacing, kneading, howling constantly. Now, even though his mistress is home, he behaves in the same way. I don't want to continue tranquilizers. What do you suggest?

A. When Siamese cats are left alone for various periods of time by their owners, they can develop abnormalities of behavior. And the presence of another companion animal does not always fill the vacuum. Seuss obviously developed a special anxiety. Sorry, I am not surprised that the cat is still having problems, even though its mistress is home. Once such symptoms are established, the pattern tends to persist. All that I can suggest is "shock" treatments. Simply place the cat in a boarding hospital for four or five days, allowing no contact with its mistress. Then let her take the cat home for two or three days and return it to the boarding facility for another couple of days. Alternatively, get the cat a companion kitten of the opposite sex for company, which is probably what he craves in his owner's absence.

Q. My cat, Marianne, has been itching since our son left home to go to boarding school. Can an emotional upset cause a skin problem in a cat?

A. Yes. As in humans, an emotional disturbance in a cat or dog may cause sudden shedding, alopecia (baldness), urticaria (hivelike reactions) and itching. Treatment with antihistamines that have a strong sedative effect is helpful, but, if possible, the root of the emotional problem should be handled, since treating the symptoms only is of little use. In your case, Marianne may need a companion.

Q. On my return from a month's trip, I found that our cat had licked almost all the fur off her tail and hindquarters. One vet prescribed phenobarbital, but it didn't help. A second vet suggests the problem is hormonal and that spaying is in order. I don't know what to do!

A. Emotional stress, such as your cat experienced when she was left alone, could well have triggered her self-mutilating behavior. If this is the case, the habit must be broken. One way is to have your vet fix an Elizabethan collar—like a lamp shade—around your cat's neck, for three to four weeks. Your vet should also examine her anal glands; if they are inflamed, the irritation might be responsible. If neither this nor the collar provides an answer, spaying or hormone replacement with a progesterone medication may be the solution.

Q. We have an eight-year-old Siamese cat. Last time we went on a trip, she got neurotic and started pulling her fur out. She was left with a member of the family, who took good care of her. This fall we are going on another trip. We are leaving her with friends. What can they do if she starts pulling her fur out?

A. This is a common behavior problem in sensitive Siamese cats. Since your cat reacts in this manner when you're away, you can try two remedies: (1) leave her at home and have your friends come there to cat-sit. She would be less disturbed at home than in a strange house; (2) your vet should prescribe a mild tranquilizer that your friends can give her if she starts to mutilate herself. Having two or three of your unwashed sweaters or T-shirts around may comfort her. Your scent would give her a feeling of security.

Q. We have recently moved to Chicago from Connecticut, and several months ago, when my parents came to visit, they brought our white cat with them. She seems to have adjusted to staying indoors, but we are experiencing some disturbing problems. She seems to have forgotten that she is litter-trained and several times a week goes on our white shag rug. I am wondering if this problem stems from moving, or from bringing inside an animal that used to have the run of the yard.

A. Any emotional disturbance, particularly a change of environment, can upset a cat's toilet routine. You can reestablish your cat's routine by confining her to one room, so that she is more or less obliged to use, say, two, three or even four litter trays that are available in it. I am reminded of the cat that became unhousebroken following the marriage of its mistress. The treatment was to rearrange one room as it was before its owner was married. The cat settled down in that room and all was well!

Q. About three months ago, I brought home a two-month-old kitten to be a companion to my year-old cat. Both cats were spayed, and in a few weeks they were apparently well-adjusted. However, in a short time one of the cats (I don't know which one) started using the dining room as a litter box, first for wetting only, and after a few weeks, for droppings, too. I took them to the vet and found that one cat had a urinary infection, which was cleared up with treatment. But the problem has persisted, and I am now quite sure that it is the new cat that is using the carpet. Do I have to get rid of her?

A. I am glad that you had your cats checked out by your veter-

The question of whether or not pet cats should be allowed outdoors will always be argued by those who take opposing sides. The owner of this wandering tomcat may feel he is doing his pet a favor by giving the cat access to the outdoors. Actually he is exposing the cat to disease and danger and allowing him to father numerous unwanted kittens and kill a variety of birds and small animals. *Photo courtesy Mary Pyles*

inarian, because sudden loss of housebreaking can be due to some disease or infection of the urinary tract. Very often, however, it is due to some change in the relationships in the household, perhaps between the pet and its owner or between two or more resident pets.

One habit of cats is to leave their marks all over the place, unburied, as a way of marking their territory. This is done when they live in the wild, and it is quite possible that some subtle conflict exists between your cats that has led to this territory-marking behavior. It is a sign of emotional insecurity. Provide the cats with two or more extra litter trays for a while and keep them as clean as possible. Also, confining the cat who messes the rug in a small room (say, the bathroom) with food, water, a blanket to lie on and litter box for three to four weeks often works well, breaking the cat's habit of evacuating in other parts of the home. Take the cat out of isolation as often as you can, and only under constant supervision, to stop her from evacuating indiscriminately.

Q. Our female cat has "told" us for years when she has a problem. She squats in front of me and urinates a small amount on the floor. And usually there is a spot of blood in the urine. When she does this, I give her Uritabs (which I get from her vet) until the problem clears up. She has been doing this since she was two years old. And she was fifteen years old last July! Our male Siamese did the same thing when he had a threatened blockage. Of course, if they had been outdoor cats, I would have missed the signals. But they do try to let us know when something is wrong.

A. Thank you for sharing with us your valuable observations. Yes, our pets aren't dumb, and if we only take the time to observe them and let them communicate with us, we realize soon just how intelligent and aware they are—and that maybe we are the dumb ones!

Q. Our fifteen-year-old Siamese sorely misses his twin brother who died recently. His howling during the night is really playing havoc with our sleep. I am not too anxious to get a kitten. Do you have any other suggestions?

A. No, I really do not. All of us—people and other animals, too—have to live through the hardship and grief of losing a loved one. Let your cat go through his natural phase of mourning, and give him plenty of supportive love and reassurance. Just as with people, when a lifelong partner dies, the companion animal may die soon after. Of course, this is not always the case. If your cat were ten years younger I would indeed suggest getting him a new companion kitten, but considering his age, he might feel somewhat put out. So let things be, and I hope he will recover soon from his sad loss.

Q. My three-year-old neutered cat has developed a peculiar condition. Every morning and evening, he seems to vibrate—down his back and in his front legs. This lasts for about five minutes. Then he's fine again, and he plays, eats and sleeps as always. Can you advise me what this might be?

A. If your cat were a dog, it would be keeling over onto one side and giving you conniptions! Your cat has epilepsy. Unlike the canine form, feline epilepsy is much more discreet. Cats just don't develop the grand mal seizures common in dogs and humans. It's some idiosyncrasy in their nervous system. Feline epilepsy tends to occur, however, in a pattern common to dog and man, namely, during the waking period and before going to sleep. If your cat doesn't develop more serious symptoms, I would leave it be and avoid medication. However, if you see that your pet is greatly distressed, there are medications available that your vet can prescribe.

Q. Every few months our cat gets a spell of hyperactivity and agitation, jumping from chair to table, crying, flicking his ears, rolling from side to side on the kitchen floor, running to and from his food, spraying in front of the refrigerator—all at a hysterical pace. Otherwise he is a sweet, lovable cat. Our vet gave him all kinds of tests and X rays and put him on Ovaban. But after five months, he had another "attack." Now another vet has prescribed one-quarter grain phenobarbitol. Please, what is your opinion?

A. Such bouts of hyperactivity in cats, in both males and females, seem to be on the increase today. Some cats are so hyper, they seem to be having some form of seizure, like epilepsy. It may be related to some hormonal imbalance, so a little progesterone therapy, like the Ovaban your vet prescribed, helps some cases. Others need sedation, and the vet who is putting your cat on phenobarbitol is taking this next logical therapeutic step. This barbiturate should help your cat.

I would also try him on a more natural "organic" diet for a while. Artificial dyes and food preservatives in commercial foods are suspected by some authorities as being one factor linked with hyperactivity in pets (and children). Purely hyperkinetic children and dogs do respond well to a stimulant like amphetamine or caffeine, a paradoxical effect. You may want to try this next: a little strong milky coffee three times a day for your cat. It may do the trick!

Q. My cat, Rocky, seems to enjoy eating clothing—not just chewing it, but actually swallowing it—and he never seems to get sick! He eats T-shirts, socks, sweaters and an occasional towel, just to name a few things. I've used all kinds of sprays on the market but this doesn't work. One time I found him eating a shirt while I was still in it.

A. Eating wool and other materials is a common vice in cats, and discipline doesn't cure it. Some cats do it because they are bored, some because they need roughage, and others because they just get "turned on" like an addict. Try feeding your cat fresh grass as a substitute. Second, figure out when he's most likely to indulge his vice. Is it when he is getting petted? Before he goes to sleep? Then is the time to provide him with his own blanket to nurse on and chomp at. After that, you should gradually wean him off his blanket until he is "maintaining" on just a few glorious minutes a day. This will stop possible complications with an intestinal blockage if he eats too much all at once. Meanwhile, keep your own clothing stashed away in a safe place.

Q. Our fifteen-month-old marmalade female cat has a persistent, damaging and potentially dangerous habit of chewing on and biting through electrical appliance cords. In addition she has chewed off the entire lower trim and casing of the kitchen door. We needed an electrician to replace the chewed through turbo-oven cord with a rigid metal-encased cord. Until a lamp, TV set, telephone or appliance fails to function, it is difficult to detect her chewing choice. She does not chew in our presence. We would appreciate your immediate response to our serious problem before she electrocutes herself!

A. Since your cat doesn't chew the electrical extension cords when you're around, there's no way of disciplining her when she's doing the job. You could sheath some of the extension cords in a hard tubing—plastic garden hose or some similar material. Also give her some extension cord pieces to play with as well as some other toys like a catnip "mouse" and a "cat mobile" (a hanging ball on a length of electrical cord fixed to a flexible wire stand).

The chewing vice you described is not uncommon, especially in

Where cats are concerned, there are no hard and fast rules. Some cats will go their whole lives without destroying anything in the household while others must be protected from their own bent toward mayhem. *Photo courtesy Mary Pyles*

Siamese cats. A very bitter solution of quinine or hot Tabasco sauce painted onto some surfaces is an effective deterrent for some cats, and is certainly worth a try. If all else fails, be sure to unplug all fixtures when they are not in use! Remember, your cat is probably bored, so your best solution may well be to get another cat.

CAT-POURRI

Finally, a few bizarre queries that say more about cats than the answers do—namely, that we still have much to learn about the inscrutable feline psyche!

Q. Do animals go off alone to die, or this is an old wives' tale? Our cat of twenty-four years (!) asked to go out early one morning and never came back. She was deaf and rather feeble, and she couldn't fend for herself. We searched the woods, but couldn't find her. What do you think happened?

A. A sick animal will often seek a safe place to lie up until it gets better. This is why animals give the impression that they're seeking solitude before they die. Actually, whether or not an animal knows it's going to die is a moot question. But we do have precedents of a terminally ill animal or one hit by a car dragging itself off somewhere to die.

All pet owners, therefore, should be watchful if they have a sick or injured animal, that it doesn't attempt to escape and find a "safe" place to rest. Because, of course, it's much safer to have veterinary help; and the chances of recuperation might be slim—or even nil—if you can't find the pet!

Q. I have been playing the piano for almost ten years, and neither of my two cats seemed to mind. But now that I'm practicing flute, the older one, Cookie, howls. Can you tell me why?

A. I haven't heard of cats "howling" to music before; sure you don't have a dog? Perhaps your flute sounds like caterwauling (no offense intended), so Cookie wants to join in. Certain sounds have been shown to have a dramatic effect on cat behavior. You may well find that it's a particular note or key that turns your cat on. If Cookie mimics the note, then the chances are she's responding socially to your flute.

Q. Our male cat goes crazy whenever someone in our family washes his or her hands with a particular cleanser. He'll come running as soon as he smells it. He'll roll all over our hands and nibble on our skin while rubbing his fur on our hands. Our other cat pays no attention to the cleanser. What's going on?

A. Cats do seem to get high on all kinds of common household items: plastic bags, mouthwash, ointments, cleansers and other household chemicals and cosmetics. Possibly, some ingredient in the cleanser mimics one of the pheromones or body chemicals cats emit as social and sometimes sexual attractants. Or there may be farm animal–derived ingredients like collagen and amniotic fluid they find attractive. Here is a warning to pet owners: Don't use any household chemical your cat seems to like. It may be licking a disinfectant spray or floor polish that is poisonous.

Q. Can you explain how my cat seems to know when I'm sick or feeling low? Also, how does he know when it's the weekend? He never wakes me early for his breakfast at that time.

A. Little is known yet about the extrasensory abilities of animals and people. It may be an ability that we have almost lost but that many animals and so-called "primitive" people are able to use. Empathy and intuition, I believe, are important keys to understanding and communion. I have "felt" and physically experienced specific illnesses in others without looking objectively at their demeanor or overt behavior. Your cat may be tuning in to you in this way, and certainly your changes in mood and behavior would be important clues.

As for your cat's awareness of weekends, animals do have an amazingly accurate sense of time, not so much by the clock, but in terms of rhythms, to which they become attuned. They have an internal timing device that sets up a circadian or diurnal rhythm. I doubt that your cat counts off the days, and think rather that he is attuned to a rhythmic cycle or periodicity of five weekdays with a two-day weekend interval.

Q. I have a funny and very unusual question to ask you. My cat does somersaults in front of closed doors, and sometimes even in front of open ones! Can you explain this strange behavior?

A. I have seen rats climb gangplanks, hoist flags and fire cannons, and I have seen pigeons actually playing football! It is possible to make animals do all kinds of extraordinary things by reinforcing certain behavior. My guess is that you have reinforced your cat's penchant for doing somersaults by allowing it to go outside immediately afterward, or providing it with some comparable reward. This response on your part caused the behavior to become more and more ingrained in your cat.

Many cats will meow at their owner and some even reach up to claw at the door. The cat would soon stop doing this if it were ignored and the door never opened! But the behavior can become so conditioned and ingrained that the cat will behave the same way when the door is wide open!

Q. No matter what anyone says, I'm still convinced that some animals understand human language. For example, when I tell my cat, Sascha, how beautiful she is, she rolls over and looks at me, as though she knows what I said, and purrs back at me, too. Explain that if you can!

A. Animals do have a comprehension of some human words. A well-trained dog, for example, can learn forty or more word signals, but this does not mean he understands language as such—verbalizations.

Most animals respond to the emotional coloring that we put around words; in a sense, this is a preverbal kind of language. The inflections, intonations and effect that we put into our voices when we talk to our pets and to each other are important elements of communication. In essence, it is not so much what we say but rather how we say it. When you are talking gently to your cat, you are in effect stroking it with your voice.

Q. When my Siamese cat wants to come in, she doesn't meow, like other cats, but instead raps on the door. It's very frightening when I'm alone in the house. How can I get her to stop?

A. This is your cat's way of communicating to you, and I suggest that you don't try to stop it, even though it does frighten you. Not all cats will cry to be let in. You will either have to live with your fear or live without your Siamese cat. And I, for one, would sooner cope with the former so I could enjoy the latter!

Q. When I first got a little cat, he never made a sound, and I thought he might not have a voice. After six years, I learned otherwise! When big cat went away to the vet for ten days, little cat began talking

Cats adopt a variety of behaviors simply by conditioning and observational learning. One of the best examples of this is the speed with which cats can learn how they can get out of the house and where to wait for those coming home. *Photo courtesy Mary Pyles*

Talk to a cat in a friendly, soothing or complimentary voice and you may be invited to a game or a petting session. Your tone of voice sends interpretive clues to the cat, and the signals it sends back can be hard to pass up. *Photo courtesy Mary Pyles*

and hasn't stopped since. They'd never been separated for more than a day or two before that.

A. Thanks for your informative letter giving us a little more insight into cat behavior. One cat may indeed rule the roost and so constantly take the initiative that a subordinate companion cat just won't talk, meow, chatter or purr! I just wonder how many socially repressed and developmentally retarded cats exist as a result!

Q. I have been away from my two cats for three months, and expect to be away from them for a considerable length of time. The person they are with is taking very good care of them. However, the cats still seem (I'm told) to be aware of my absence. What can be done to make them remember me when I do come home?

A. That's hard to predict. It depends on how old your cats are and how long they lived with you and were imprinted. A strong imprint is hard to break. When you come home after a long absence, your cats may not recognize you at first. (And remember, any strange odor on you could freak them out initially.) During your absence, it would be a good idea to occasionally send your cats an old T-shirt, unwashed, and smelling of you. Also send them a ''smell'' letter in the form of a piece of paper you have rubbed on your forehead, hands, arms and lightly under your armpits. This may sound rather peculiar, but it should work because you will be maintaining contact with your cats.

Q. Am I right or crazy? I am convinced my cat is psychic. He knows when I'm sick or depressed and seems to know what I'm thinking. For instance, the other day as soon as I decided to fix myself a tuna sandwich, he was by the fridge meowing. I hadn't said a word, and it was way before his dinner time.

A. Cats are intelligent. Intelligent, sensitive, alert and empathetic enough to be psychic, to have ESP? Yes, I think so, but only cats can prove it to scientists, who do not necessarily believe (for that can bias the results) or disbelieve (since such negativity will disrupt the cat's abilities). So where can we find a scientist who is objective, totally unbiased? I don't think I can, for such a person would perhaps not be a scientist who would bother to prove the obvious that can only be appreciated, understood and *actually experienced* by believers. So be it. Such are the intrinsic limitations of science today. We need faith, perhaps, and certainly empathy more than we need ''hard facts''!

Index